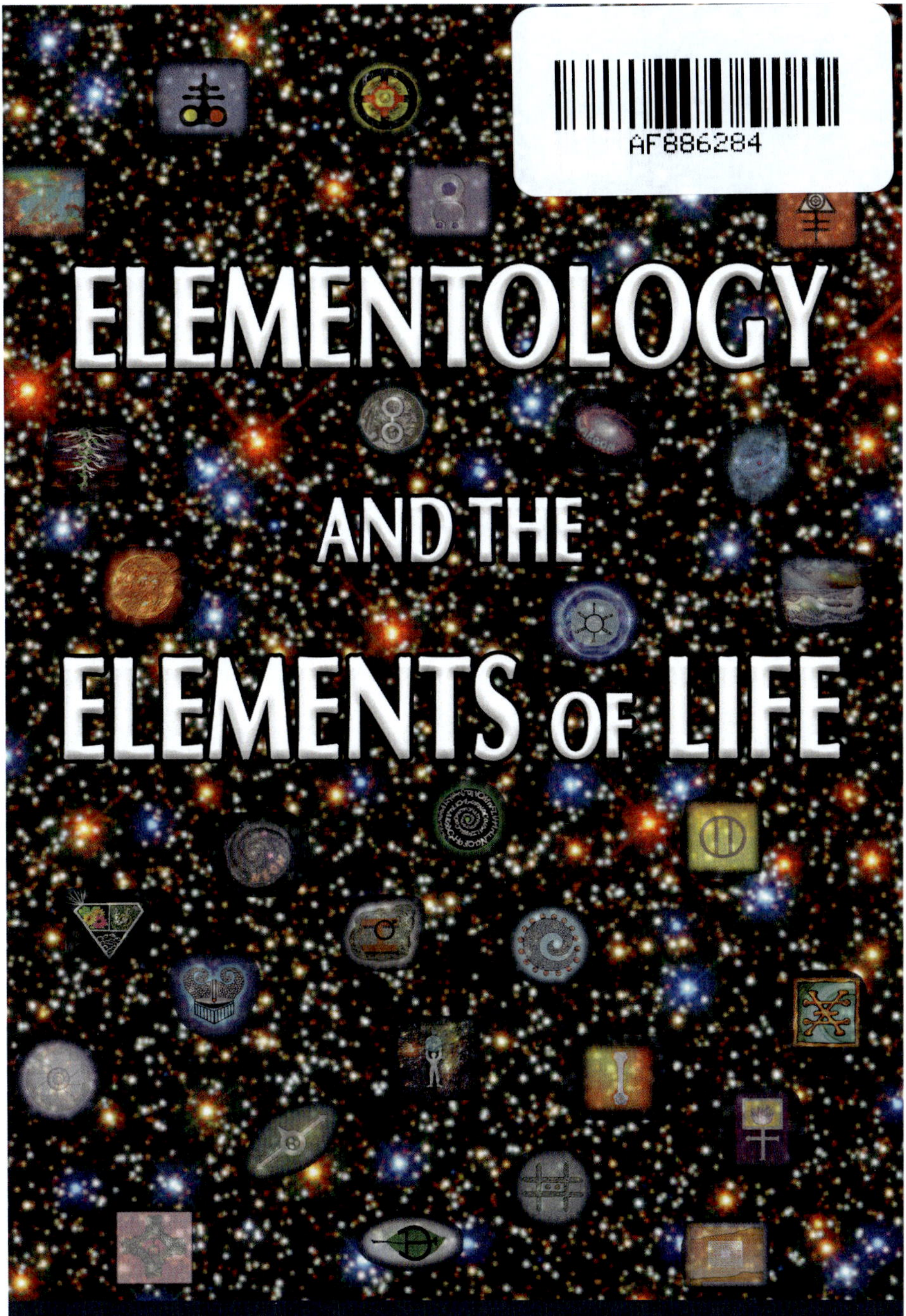

ELEMENTOLOGY AND THE ELEMENTS OF LIFE

By Patrick Collins, P.E.

ELEMENTOLOGY AND THE ELEMENTS OF LIFE

By Patrick Collins, P.E.

To purchase additional copies visit
www.elementologybook.com

ISBN-13: 978-0-9975011-1-7

Cover and Book design by Sandra Smith Larson

Published by
EXCEL AVENUE PRESS

EXCEL
AVENUE
PRESS

CONTENTS

ABOUT THE AUTHOR

Patrick Collins has had a relentless interest in science since early childhood growing up in Pennsylvania. Derived from his lifelong passion for, knowledge of, and experience with science; and coupled with his obsession of the universe, our existence, and what we are, Patrick has created the revolutionary field called Elementology™.

Patrick holds a degree in engineering. He was inducted into the National Physics Honor Society and holds numerous Professional Engineering Licenses.

Patrick has resided in Atlanta, GA with his teenage son P.J. since 2000.

Elementology™ is an amazing new field that is derived from the science of the elements. Given that we are completely comprised of elements, Elementology™ proposes that we must have some governing principles that we share with them. It introduces concepts that provide insight to our personality characteristics, and behaviors.

DEDICATION

This book is dedicated to my mother and
late father for providing the elements required
to make my existence possible.

SPECIAL THANKS TO:

Mark Siegel - thank you for your encouragement,
support and all of your help.

Chris Collins - thank you for your support,
ideas, time and for being a wonderful friend and brother.

Clark Davidson - thank you for your support,
friendship, and for believing.

Margaret Collins - thank you for all of the work and
effort you put into this book and throughout my life.

To my son: thank you for the inspiration you have provided me and
for being the best son a parent could ever imagine having.

FORWARD

My name is Dr. Lovve. I am one of the world's premier relationship matchmakers (DrLovve.com). Psychology is my predominant tool when matching men and women for love. However, I also utilize numerous other philosophies and sciences of human psyche and behavior to assist my matchmaking practice. My passion is to help others find love and relationships. As a result, I am always seeking new and creative predictive methodologies of human behavior, personality, and compatibility with others. I have recently been very fortunate to discover Elementology™. My staff and I were so excited and fascinated when we read *Elementology and the Elements of Life*© that we at LovveMatch immediately began to incorporate its concepts and theories into our process to help men and women find, keep, and nourish a more fulfilling love life for themselves. I personally have been enthralled with the incredible new concepts that Elementology™ unveils and have incorporated them into my own personal growth as well as my relations with others.

As a scientist, I like data. I also appreciate that science is heavily fueled by statistics. Numbers and statistics are important ingredients for reputable and reliable analysis. Statistical analysis greatly enables all of us to understand aspects of our world more appropriately through an objective, unbiased lens. Compiled data that are gathered, then analyzed strategically can be easily processed. Valuable data can provide humans with vital and relevant information. Pertinent data can literally help us revolutionize our lives for the better. For almost a decade I have collected clients' birthdates to accentuate my psychological matchmaking work with love and destiny astrological reports (LoveandDestinyReports.com). I have compared this data to the profiles detailed in *Elementology and the Elements of Life*©. The general accuracy of the profiles versus the data I have collected is astounding and is exactly why I feel so fortunate for having discovered the concepts and theories of Elementology™. Elementology™ has taken the predictability of human behavior, personality, and compatibility with others to another level. Elementology™ provides me with another piece of a given client's personality pie which enables me to more accurately predict personality characteristics, behaviors, and therefore connections with others. I have grown personally and as a scientist of love for having read and integrated Elementology™ into my personal life and my matchmaking practice.

Dr. Lovve

DrLovve.com

(888) 854.7488

INTRODUCTION

Have you ever met someone and immediately "clicked" with them for no apparent reason? Or have you ever instantly felt that you were totally adverse to someone? We've all had these immediate reactions to other people. But what is the basis for this instant interest or revulsion? It is likely a result of both inborn and environmental factors that invoke these responses. The environmental portion, or the personal experiences each of us has been exposed to throughout our lives, shapes our judgment and assessment processes of others. The other portion that governs these responses is instinctual, or a sixth sense if you will. Could these instincts and other innate characteristics that each of us possess be guided by something so basic, something so fundamental, something so … elemental?

The elements are the "building blocks" of Nature. Everything in the universe, from a microscopic molecule to the largest star, is made up of elements. Each of the known elements has certain characteristics that guides its behavior, including whether and how the element interacts with other elements. Some elements are drawn to certain other elements and combine with these elements to create molecules. These same elements however will be repelled by certain other elements and therefore will not readily bond with those elements.

All of us are completely made up of elements. Therefore, there is an undeniable connection that we have with the elements. Elementology™ is based from the scientific data of the elements. Elementology™ describes how each of us has a Life Element. Elementology™ details how many of your own innate personal traits, tendencies, and characteristics are related to your Life Element. So that instant connection you feel meeting someone new is not just dictated by your experiences and life lessons, but it is also influenced by your Life Element.

Given the fact that we are physically nothing more than a compilation of elements, this groundbreaking book unveils the practice of comparing and applying the characteristics and behaviors of the chemical elements to our own characteristics and behaviors. To this point in history we have surprisingly over-looked this undeniable relationship that we have with the elements. The elements are who and what we are, nothing more and nothing less. This book defines what Elementology™ is and how it works. It introduces the personality profiles for the 31 different Life Elements (one for each day of the month), and explains how your Life Element predicts aspects of your personality.

Each of the Life Element profiles describes the unique physical characteristics and behaviors of the element. It also illustrates where it appears in nature, and describes what humans create with it and or how we use the element. It explains how these characteristics, behaviors, and uses of each Life Element shape and predict our own characteristics, behaviors, and interests. Finally, each profile specifically identifies the other Life Elements that are most compatible, and least compatible, with your own Life Element. The profiles are brimming with interesting scientific, medical, historical, and general information. Find out what your Life Element is, and determine if you are optimizing your life to provide the most positive outcomes that you can. Learn about your Life Element, who you are more likely to get along with, and who you may be better off avoiding. Compare your friends, family members, and significant others' Life Element to see if you are compatible and what compound your two elements might combine to form.

Elementology and the Elements of Life© takes you back through time to consider mankind's never-ending quest to make sense of the universe, to determine where we came from, and to figure out who and what we are. It will challenge you to consider and ponder some concepts that vary from the traditional and long-held teachings and perceptions of the universe. This book also discusses the vital role that astrology/astronomy has played in the survival and advancement of our species. *Elementology and the Elements of Life*© is certain to captivate your interest and have you reevaluating the traditional teachings and perceptions of our existence, your inner workings, and your relations with others.

THE EVOLUTION

Since the beginning, humans have grappled with the Big Questions: "Who are we?" and "Where did we come from?" Even today, we continue to ponder our connection with the universe and the vastness of its planets, stars, and galaxies. We dream about the possibility of other life forms, other "states" of energy, time, space, and matter, and perhaps even other "universes".

Astrology is the first known effort that humans made to connect our lives with the heavens. Astrology has had a long evolution throughout the entire known recorded history of mankind. Modern western astrology that is practiced today is a method of predicting events or insights into one's personality based on the positioning of celestial bodies (such as the Sun, the moon, planets, stars, and constellations). Today astrology is considered to be both an art and a science. Some people dismiss it altogether and find no worth in it. Others find truth and value in it and incorporate it into their daily lives. Still others view it as interestingly coincidental or find it to be entertaining in some way. Regardless of where you fall on this spectrum of believers, non-believers, and casual amusement seekers, the history of astrology and its contribution to perpetuating the existence and the advancement of human civilization is stunning and undeniable. Most of us are unaware of the vital role that astrology has played in the survival, progression, and advancement of human kind. Astonishing to most, without it we very likely would not be here today.

Astrology and astronomy were the same science for thousands of years. Based on ancient cave carvings, astrology/astronomy is documented to have been practiced as far back as at least 25,000 years ago to record and predict weather patterns based off of the cycles of the moon. As time progressed, human knowledge of the heavenly bodies and how they relate to the Earth kept expanding. This led to deeper understandings of how seasonal activities were linked beyond just the moon, but also with planets and constellations which appear to us in the sky. Even our ancient primitive ancestors realized that our seasons change in harmony with the changing positions of the Sun, moon, planets, and constellations. Therefore, we concluded that there must be a connection between the positioning of the bodies in the universe and events on our planet. As our species progressed further and gained even more knowledge of the relationship between the heavens and the Earth, we developed the first astrological calendars. These calendars detailed seasonal tasks such as harvesting, hunting, fishing, managing water reserves, and other tasks related to survival. Astrology/astronomy and these astrological calendars have played a vital and remarkable role in both the survival and progression of our species.

Many ancient cultures practiced astrology in varying forms to help predict weather and seasonal events which ensured their survival. The Babylonians were the first to have an organized system sometime around 2,000 BC. This was also the first known form of astrology to include mathematics. This system went beyond just predicting the seasons, weather and survival events. The Babylonians used astrology to predict the outcome of political matters, the chances of success in waging wars, and other events and happenings.

Following the conquest of Egypt by Alexander the Great around the 3rd Century BC, Babylonian astrology began to merge with the Egyptian tradition of Decanic astrology. This combination of Babylonian and Decanic astrology gave birth to Horoscopic astrology. The revered scientist and astrologer/astronomer Ptolemy is credited with compiling these prevailing astrological methods of the era. He combined these ancient astrology methods with his scientific studies and theories of celestial bodies, which together provided the fundamental basis for both western astrology and the science of celestial mechanics. Ptolemy's work in turn, inspired other great minds from Copernicus to Galileo to Kepler to Newton to Einstein and on... Ptolemy's ideas survive to this day in both modern science and western astrology. He documented these concepts in a collection of writings titled "Tetrabiblos" in the 2nd Century AD. This work was considered the "bible" among astronomers/astrologers for an astonishing 1,500 years! Though some of Ptolemy's scientific theories have been proven inaccurate, many are still valid and utilized today. Ptolemy's ideas and scientific advancements helped shape our understanding of the movements and relationships of heavenly bodies, and thus astrology as well. His contributions to science in general, astronomy, and modern astrology are fundamental.

WHERE DID WE COME FROM AND WHAT ARE WE

Where did we come from? What are we? Envision being one of the earliest modern humans on the planet. With such little knowledge of science and of what transpired before us, how would we perceive and interpret the world? We can only begin to imagine what thoughts and ideas filled their minds regarding these and other questions. Since these early times, humans have sought to understand how we came to be; what we are made of; what our planet is and how it "operates"; how do we fit in to the "grand scheme" of it all; how the Earth related to the Sun, the Moon, and the rest of the stars in the night sky. It is natural for humans to attempt to make sense of their surroundings, to try to understand how we came to be, and to make order of the randomness and chaos in which we exist. As humans have evolved we gained greater knowledge and a better understanding of the universe and our place within it. However, even with all that we have learned, we have made little progress to answer the Big Question: "Where did it all begin for us?" While humans continue to struggle with that question, we have concluded what we are.......we are elements.

The consensus among scientists is that "all" the known matter (elements) in the universe is confined to a roughly 13.8 billion light year range which is conveniently observable to humans. That is, we can only detect elements as far as a span of approximately 13.8 billion light years. All of this matter was created at the time of the Big Bang from a single point and has been moving outward from that point for roughly 13.8 billion years. There is little dispute about the existence of this 13.8 billion light year range of the elements, and that it derived from a single point (as crazy of a concept that is to imagine), as there is much evidence to support this theory. However, most scientists speak of this matter in the 13.8 billion light year range as the "universe" and that all the elements within this range are "all" that has ever existed. Is it logical for us to confine our universe to such a mere amount of space and matter? And should it be reasonable for us to confine our focus and minds to such a limited range?

“Mere” is said in a sort of satirical sense as the range of 13.8 billion light years is enormous……….or is it, relatively speaking?

Is it more plausible to believe the universe goes beyond the commonly accepted approach that all the elements came from a single point - aka the “Big Bang”? Again, based on what we know at this time, there is little disputing the Big Bang. However, is it reasonable to buy into the common belief that this is the “only point” or “place” ever from which ***all*** matter/elements were derived from? Who is to say that there have not been or that there may be an endless amount of “points” throughout the universe that have created or will create other “Big Bangs”? Based on what we reasonably know at this time, it is safe to say that the very elements that collectively make up your body, your brain, and everything on Earth derived from that single Big Bang point that we are aware of. However, it does not even address, never mind does it answer the questions if these are all the elements that exist at this time, or if they are all the elements that have ever existed, or that will exist. It also does not answer what these elements were, or where they came from prior to the Big Bang.

There is a phenomenon that has occurred throughout our scientific history that continues to perpetuate through time, even to this day, that relates to our understanding of our existence, our place, and our conclusions of these things. For example, we once believed the Earth was the center of the universe; later we concluded our Sun was the center of the universe; now most scientists argue that the universe and its contents (all the matter/elements we see) came from a single point - the Big Bang. If we consider these false conclusions throughout history, shouldn’t we be more cautious and consider that the plot will likely thicken? And therefore, that from where we conclude we came from at any one given time in history will also likely continue to evolve. What we believe regarding the center of “everything” and or “creation” has only led us to something more each time….

Let’s take a moment and envision the universe from a different perspective than the commonly conceived point of view. In order to do this we must open our minds and put aside all traditional teachings and beliefs for the moment… Are you ready? OK! Let’s picture the trillion or more other galaxies that surround us in the 13.8 billion light year range of matter that we are aware of as one "grouping". Envision this "grouping" to be just one very tiny piece of matter in a much larger "universe". Imagine that there are an endless amount of other "groupings" of matter like our 13.8 billion light year range that are spaced out hundreds of billions or even millions of trillions of light years away from one another. Picture these other "groupings" of matter either being similar in size (13.8 billion light years range/with trillions of galaxies), smaller, or bigger. In this massive scaled universe, let's also envision our entire 13.8 billion light year range of matter as just the matter that makes up a tiny proton within an atom in this large scaled universe. That is, our entire 13.8 billion light years of matter is actually just a tiny part of, what we would consider from our human perspective, an unimaginably colossal sized atom. That would mean that everything that we know of (all the matter within the 13.8 billion light year range) is really just a tiny fraction of the matter that exists within a massive universe.

Let's also view time from a different perspective too. It has taken 13.8 billion years for the known matter in our universe to disperse to where it is now. Imagine that timeline being incredibly short in the relative length of time that it has existed (13.8 billion years). Imagine all that has occurred dating back 13.8 billion years to now as just a fraction of a second relative to the mega scaled universe we just imagined. What we conceive as either microscopically small or colossal regarding either the size of matter or length of time is really just a context of perception. This is a very important concept when one views our universe and our existence. Our ability to imagine different perspectives are limited when we confine our minds to what is easy and convenient to understand. We further limit our ability to consider other viewpoints and thus may fail to make sound conclusions when we view things from an egocentric position. This is true especially as it relates to time, space, and matter as we as humans perceive it. We do not know with any certainty whatsoever how much matter is in the "universe", or how big each "grouping" of matter can be, or how many "groupings" of matter there are. We also have no idea how small matter can get. It could be infinite on both ends of the big and small spectrum - what a crazy concept to ponder! Could the massive universe we just imagined be just a fraction of an even larger universe and on and on to infinity? And could atoms be broken down to smaller components; and those components be broken down, and on and on to infinity? With all that said, it seems rather shortsighted to conclude that there is only 13.8 billion light years of matter just because we humans are only capable of detecting only that matter at this time. Everything is relative to scale whether it be time, matter, or space. We must remove ourselves from the human viewpoint as much as possible when we consider space, time, and the abundance and existence of matter. This is an important and key concept when deliberating the question "Where did we come from?"

So, is it an easy answer and convenient for us to simply think that all the matter we are aware of at this time is all that there is or ever was? Similar to how convenient it was to settle on the belief that the Earth was the center of the universe, then the Sun, and so on. Should we consider what history has taught us? Our judgments and conclusions throughout our history have been and will remain as limited as the knowledge and capabilities we possess at any given time. Is it proper to limit our focus and perception to the commonly accepted theories? Or should we be far more inclined to believe that there is likely to be so much more to the universe than the mere 13.8 billion light years range of matter/elements that we can detect at this time? Shouldn't we be open to the possibility of the 13.8 billion light years range, that we do see and can say for certain exists, as just a tiny infinitesimal fractional drop of matter in an endless sized bucket of matter in an endless space? Perhaps we are not capable of understanding this, or viewing additional matter/elements at this time beyond what we can see, detect, or observe for what could be endless reasons. Perhaps matter exists beyond some other space "medium" which we are incapable of observing, "seeing" through, detecting, comprehending or understanding.

There are many mysteries to be uncovered in our search to answer the question; "Where did we come from?" However, what we do know with some confidence is what elements exist within the 13.8 billion light year range of known matter. And with even more certainty we know what elements comprise our Earth and all of its contents including our brains and our bodies. It is undeniable that the elements are what and who we are and therefore, the elements shall dictate who and what we are....

AN INTRODUCTION TO ELEMENTOLOGY AND THE ELEMENTS OF LIFE©

The elements make up all planets, moons, stars, and galaxies within our universe as we know it to be. The elements also make up all human beings as well as all other living and non-living things on our planet. The elements define who and what we are. We are matter (elements) and all that surrounds us is matter. Thus, we and our reality do not exist without the elements. What it all comes down to is, we and all other matter, are nothing more than a collection of elements. All that we are, and all that we possess on this Earth are nothing more than the compilation of either a pure element, a natural compound made up of elements, or a compound created by man manipulating the elements. Given that we are wholly comprised of elements, there is an undeniable connection between us and the elements - the elements are who and what we are. For too long, we have overlooked this fundamental relationship between ourselves and the elements.

In many ways, our behaviors are guided by and related to principles that dictate the behavior of the elements. For example, most elements are innately driven to "react" with, "combine" with, and "bond" with other elements. Most elements are attracted to other elements, they do not like to be alone. Similarly, most people are driven to "react" with, "combine" with, and "bond" with other people - we too tend to not like to be alone.

Each element possesses unique physical characteristics which dictate its "personality". These characteristics provide predictable patterns and behaviors. Each of the elements responds uniquely and with certainty to their environment. The elements are also predictable in the manner of how they interact and bond with other elements. Elementology™ compares the characteristics and behaviors of the elements with our own characteristics and behaviors. Elementology™ explores the analogous relationship between the characteristics, tendencies, and the interactions among elements to our own characteristics, tendencies, and interactions with others. Our personalities and traits, our day to day compatibility with others, our friendships, and romantic relationships are correlated to our Life Element. This comparison process provides us with insights into our innate qualities and who we are better suited to interact with.

The way each of the elements behaves uniquely is remarkable. The behavioral characteristics of each element are determined by the number of protons in its nucleus (center). For example, every Oxygen atom will always have 8 protons - that is what makes it Oxygen. Protons define what each individual element is, and provides each element with its particular characteristics and behaviors, including how and what other elements it reacts with. The number of protons in a nucleus inexplicably provides the unique characteristics and behaviors for each element.

The day of the month we are born on corresponds to the element with the same atomic number (the number of protons in its center). For example, if you were born on the 8th day of a given month your Life Element is Oxygen. Oxygen has eight protons in its nucleus and thus its atomic number is eight. If you were born on October 8th you are an "Oxygen-Libra". Thus, you would possess the characteristics of both Oxygen and Libra profiles based on Elementology™ and astrology. Because Libra and Oxygen are both air signs, Oxygen-Libra's are compounded Air signs.

It would be reckless to say that "who we are" is limited to our inborn qualities only. That is, it is not that easy to just simply say we are what our Zodiac Sign, Life Element, or any other predictor of innate characteristics tells us what we are. We are also greatly influenced by many factors in the environment that we are exposed to throughout our lives. Our innate characteristics and qualities stay with us, but these inborn qualities are often altered, suppressed, or exaggerated in some ways by our life experiences. Imbalances from toxic relations can arise from being involved with people we should not be with; whether they be lovers, friends, business associates, etc. Mental instabilities brought on by numerous factors such as prolonged stress, exposure to toxins, child/emotional/physical abuse, and drugs/alcohol can all wreak havoc on our given traits. Chemical imbalances, diseases, defects, mental or physical traumas, etc. can also alter our innate characteristics. All of these factors can affect our inherent character and who we truly are - they can throw us off balance from the intrinsic qualities that our Life Element dictates.

FURTHER UNDERSTANDING ELEMENTOLOGY AND THE ELEMENTS OF LIFE©

Elementology™ is based on science along with an interpretation component. The profiles for each Life Element are based directly on the scientific data for each element. The interpretation portion of Elementology™ is found in the translation process of each Life Element. Many of the qualities and characteristics of the Life Elements can easily be directly translated. However, each Life Element has many characteristics, behaviors, uses, and some or many of these characteristics and behaviors do not always complement one another. This makes translating the personality profiles a bit complex as multiple differing characteristics and behaviors must be weighed to conclude specific qualities of each Life Elements personality profile.

The Elements of Life™ are the first 31 elements in the chemistry Periodic Table. They are the elements that possess 1-31 protons in their nucleus (center). Hydrogen has 1 proton in its nucleus and thus is associated with the 1st day of the month; Helium has 2 protons in its nucleus and is associated with the 2nd day of the month; Lithium has 3 protons in its nucleus and is associated with the 3rd day of the month; and so on.

Some of the categories of qualities utilized to determine the individual Life Element profiles include but are not limited to the following: physical state and physical characteristics of the element at standard conditions; electronegativity (how reactive is it with other Life Elements); how does it form bonds with other Life Elements (does it give up, take or share electrons when bonding); does it typically react calmly or violently with other elements; type of compounds it creates with other elements; the occurrence/presence of the element on Earth and is it required in life forms; what naturally occurring compounds on earth is it present in; what man-made compounds is it present in; uses of the element (in nature and by man); reactivity with water, bases, acids, and weathering characteristics; unique attributes; and heat and electrical conductivity.

The terms soft, hard; smooth, rough; strong, weak; simple, complex; stable, unstable along with many other adjectives are used to describe the properties and characteristics of the elements. These same adjectives used to define the characteristics of specific elements are also used to describe the personality characteristics of people. The same qualities that a Life Element possesses will be present in those people that are born on that Life Elements day. That is, the unique characteristics of a specific Life Element will translate to the innate qualities of individuals that possess that Life Element.

The term *standard conditions* is a phrase that refers to a set temperature and pressure. Each of the Elements of Life exists in different phases at different temperatures and pressures. The four states they exist in at *standard conditions* are: solid (Earth signs), liquid (Water signs), gas (Air signs), and if flammable/ explosive (Fire signs). Depending on the temperature and pressure, each of the elements will exist as a solid, liquid, gas or even be explosive or spontaneously ignite or be flammable. For our purposes we use average room temperature and pressure that we experience in a temperature controlled home or office as *standard conditions* to provide the base phase of each Life Element.

Electronegativity is perhaps the most important quality that governs critical behaviors of an element and will therefore also dictate crucial parts of one's personality. The major areas that electronegativity controls are the reactiveness one may have with others, how social they are, and how one balances their instincts/emotions with their thoughts/analysis. There are other contributing factors that also play into how reactive/social one is and also how one balances their instincts/emotions with the analytical portion of their minds. However, electronegativity is the major factor involved in these characteristics. All the elements have a certain electronegativity or a degree to which they interact with other Life Elements. Generally, the higher the electronegativity, the higher degree to which that element will seek out and react with other elements. Also, in general, the higher the electronegativity, the more one will rely upon their instincts/emotions versus thought/analysis.

Certain people are more outgoing, and find it easier to relate to many differing types of people and in different settings, venues, etc. The elements also behave in this manner. Let's use Neon and Fluorine as examples. Fluorine is extremely reactive (it possesses the highest electronegativity) and therefore these personality types will tend to have a higher need for socializing and being with other people; they will also likely be less picky with who they interact with. Fluorine personalities also find it easier in general to interact with many varying types of people in differing situations. In contrast, Neon personalities are more stable and non-reactive (it has one of the lowest electronegativity ratings) and therefore tend to be more innately shy, enjoy alone time more than most, and will likely be more selective with who they interact and bond with. People who have a more reactive Life Element, such as Fluorine, will tend to be innately less shy, be more outgoing and social, more passionate, more impulsive, and find it easier to make new friends to list just some qualities associated with a higher electronegativity. On the other hand, people who have a less reactive and more stable Life Element will generally tend to be more reserved, less outgoing, less reactive/passionate, enjoy and require alone time more than others, be more controlled, and may find it more difficult to relate to different types of people compared to others. However, this does not speak to the quality of interaction that each of these Life Elements may have with others. This is governed by other attributes. For example, though they are very outgoing and easily interact with most others, many reactions that the element Fluorine has with other Life Elements are violent, explosive, or result in toxic substances. Therefore, Fluorine personalities may easily interact with others, make friends, and have passionate relations, but they will also likely be more prone to drama, conflict, and some of their relations may be rather volatile even from the get go.

All of us react to and make judgments and decisions about numerous things each day. Some of these things are trivial and bear little to no impact on our lives or the lives of others that we cross paths with. However, other reactions, judgments, and decisions can impact our lives and the lives of others greatly. Sometimes we react emotionally, and at other times we are more thoughtful and calculated about our reactions, judgments, and decisions. Be that as it may, each of us has innate tendencies that dictate whether we are generally more emotional and intuitive versus thoughtful and analytical with our reactions, judgments, and decisions. In other words, some people more often lead with their hearts over their heads. While others will more often lead with their heads over their hearts. And yet others may have a very good balance of the two.

Some people are more reactive in general and don't think things through as much when compared to others. These traits have a direct correlation with the electronegativity of an individual's Life Element. There are also other qualities of a Life Element that can factor into this portion of one's personality, but generally electronegativity rules this domain over any other aspect. Let's use Neon and Fluorine again as examples. The Life Element Neon is very stable and less reactive than most. In fact, it has nearly no electronegativity. Therefore, Neon personalities are generally expected to put more thought and analysis into their reactions, judgments, and decisions over emotion and intuition. Fluorine personalities, on the other hand, being very highly reactive are far more likely to respond through intuition and emotion. There is not necessarily an advantage to being thoughtful and analytical versus relying more on emotion and intuition. Rather, it is more about being congruous with your Life Element qualities. That is, if you are a Neon but you repeatedly act on intuition and are very emotionally driven, you are out of tune with your Life Element. As a result, you may make decisions and choices that do not provide the best results for you. Further, being on the extreme of the scale, like Fluorine and Neon are, can bring on qualities that must be recognized and managed appropriately. For example, Fluorine's will more often have a tendency to have over emotional reactions, tumultuous and contentious interactions compared to others. On the other hand, Neons can sometimes be viewed as cold or distant due to what can be perceived as an aloof demeanor, lack of emotion, or over shyness at times. ***When living in harmony with one's self, one is most cognizant of both their inherent and learned characteristics. A disciplined awareness of ones governing characteristics enables one to achieve a control to execute and operate optimally in their relations with others and their intrapersonal dealings***. Another way of saying this is, if you are aware of your learned tendencies (and when interacting with others you are aware of their learned qualities too), and you are also in tune with your inborn qualities (and when interacting with others you are aware of their inherent qualities too), you can manage your strengths and weaknesses in a manner that provides the most positive outcomes for you. This is typically not an easy task as it can take considerable conscious effort and high intention by one's self. And when interacting with others (friends, family, lovers, etc.), it can also take much effort by them as well.

Some Life Elements are relatively stable and are moderately reactive with other Life Elements. Based on these qualities alone one might conclude that they would lean toward being more calm, easy going, less impulsive, have a low level of drama/conflict, possess relatively moderate patience, and be able to handle intense situations and manage stress well. However, it is not that straightforward and easy. Many stable Life Elements can have fierce reactions or produce harmful compounds with other Life Elements. These unfavorable interactions can be indicators of a Life Elements incompatibility with certain other Life Elements. It can also aid in predicting their general disposition as well. Some Life Elements in their pure form can break down readily in the environment or when exposed to just mild acids or bases - these qualities can represent an inability to manage or handle stressors in one's environment. A Life Element that is a poor conductor of heat demonstrates that their physical properties will deteriorate at a low temperature. This indicates that one may not handle pressure very well or hold up in contentious situations. We've all heard the saying, "If you can't handle the heat, stay out of the kitchen." All of these characteristics indicate a Life Elements ability to handle dire situations, manage stress, and gauge one's patience, to name a few characteristics. Additionally, electronegativity can also play a role in these areas too. Once again, using Fluorine and Neon as examples, Fluorine personalities will tend to be more prone to stress and be less patient at times as a result of their high electronegativity (this is generally the case but as mentioned, it will vary based on other factors as well). They will also tend to be involved in more tumultuous relations, and may overreact emotionally compared to others. Neon however, will trend the opposite and thus be more likely able to better manage tumultuous and dramatic issues in their lives. All of these factors and more must be considered when determining how a Life Element handles stress, confrontation, and dire situations.

Some Life Elements in their pure form will form a protective coating on their outside surface when exposed to standard conditions or weather. The degree to which they form a layer and how quickly it is formed, in conjunction with other varying characteristics and factors such as one's electronegativity, as one example, can determine how open, guarded, or protective a Life Element may be in general.

The qualities of "giving" and "sharing" are based mainly, but not totally, on the way a Life Element will bond with other Life Elements. When elements bond, they will either give up, take, or share electrons with another element(s). Some elements can do all three. That is, they can give up, take, or share electrons depending on what other elements and combination of elements they are bonding with. The Life Elements that typically or most often give up electrons are said to generally be more giving and charitable in some form. This is especially true for those that generally give up multiple electrons when bonding. This also is an indicator for how one may conduct themselves in this regard in both their one on one and romantic relationships with others. For example, those Life Elements that give up electrons when bonding will innately be more of a "giver". The Life Elements that take electrons when bonding will tend to be "takers". Additionally, the Life Elements that typically or most often share electrons when bonding will tend to enjoy sharing what they have with others and will appreciate when others do the same. They will both give and take, and gain a higher sense of connection with others through this process. Those Life Elements that also tend to share electrons with multiple other elements at the same time indicates their tendency to enjoy being with groups of people, active in clubs, or other activities that involve multiple people.

The naturally occurring and "man-made" uses of Life Elements are translated to determine what interests each Life Element will tend to be drawn to. Each Life Element has unique occurrences and combinations that are found in Nature on Earth. The elements are also used by man either in their elemental state or through manipulation to produce new compounds that are useful to us. It is interesting to put in perspective that this manipulation of the elements is essentially the major component to nearly all of the scientific advancements that we have created and use in our daily lives and beyond. The elements completely make up everything that we possess. They also make up everything that we eat and drink. Thus, we continually need to provide new elements to our bodies to live.

Certain elements are used to make specific products such as steels, plastics, textiles, inks, electronics, etc. Or they are used in processes or as catalysts to produce the aforementioned products. Or they are used by Nature for things like respiration, photosynthesis, creating DNA strands in living creatures, etc.

Let us consider a car as a typical example of a product that is comprised of many different parts and elements. All of the pieces that make up a car, every component from tires to windows to steel to plastic to the lighting to electronics, etc., are made from different elements or different combinations of the elements. Many processes and catalysts including chemical reactions to speed up or slow down elemental reactions are needed to produce the desired plastics, steels, compounds, etc. that are in the car. Some naturally occurring uses of elements include the conversion of Oxygen to Carbon Dioxide through respiration; the burning of Helium by the Sun; photosynthesis; or the thousands upon thousands of reactions between the elements in living bodies to name just some examples.

All of the specific products, processes, or natural uses or occurrences are unique to specific Life Elements. Therefore, these are characteristics that can be associated with, translated and attributable to those Life Elements and their respective profiles. For example, Magnesium is the Life Element that is at the heart of photosynthesis. Without Magnesium, photosynthesis would not exist, and there would be no plants, flowers, trees, etc. Thus, Magnesium personalities are more likely to have associations in some way with plants, trees, flowers, etc. Another example is how Iron is used to make buildings, cars, and all sorts of structures. Therefore, it is more likely that someone possessing the Life Element Iron will have interests and/or associations of some kind with architecture, construction, cars, etc. Another example is how Oxygen and Hydrogen bond to form water. These two life elements are therefore likely to have an affinity for water in some way.

Many of the Life Elements profiles share similar characteristics and interests. This is to be expected as we all share the same basic needs such as food, water, air, clothing, shelter, along with many similar recreational interests. Also, most people seek and desire the company of other people. After all, we are social creatures that seek to engage and bond with others. Similarly, the elements are very social as they "seek" the company of other elements. Therefore, it is not surprising that many of us are "outgoing" or "open". However, some of us are more outgoing and social than others. Similarly, some elements are more engaging and reactive with other elements. Additionally, even though we are all different there are many commonalities that we share - this is also true for the elements themselves. For instance, most people read, enjoy traveling, dining, etc. Therefore, it should not be surprising that the profiles for many of the Life Elements will have similar interests and that they will also share some similar personality characteristics.

The natural occurrence of Life Elements in compounds on Earth is one of the biggest factors, but not the only one, that determines who we are compatible with. Using Sodium (11) and Chlorine (17) as an example: Sodium (11) personalities (people born on the 11th day) are most compatible with Chlorine (17) personalities (people born on the 17th day) and vice versa. These two Life Elements are most compatible with one another because the combination of Sodium (11) and Chlorine (17) (Sodium Chloride) is the most frequently naturally occurring compound on Earth between these two Life Elements. Sodium Chloride (which we readily know as common table salt) is found in great abundance in the Earth's crust. It is also found in high concentrations dissolved in the oceans.

Another factor that must be considered regarding compatibility is man-made compounds. There are many man-made compounds that are essential to our daily lives. Therefore, Life Elements that combine through the help of man and are useful to us are certainly compatible. An example of this includes Iron (26) and Carbon (6). Iron (26) is a strong steel but when combined with just a little bit of Carbon (6) it drastically increases the strength and usefulness of Iron (26). Another example is Iron (26) and Zinc (30). Iron (26) is extremely susceptible to corrosion but when coated with Zinc (30) it provides a protective layer that helps preserve the Iron steel.

THE ELEMENTS OF LIFE: CONNECTIONS, RELATIONS, AND LOVE

Each Life Element profile lists all the other Life Elements that each of us will be more likely to be compatible with, and also what other Life Elements we will be less likely to be compatible with. The compatibility and incompatibility sections of each Life Element profile details what Life Elements you are more or less likely to strike a connection with, have relations with, or even fall in love with. There are a number of factors that contribute to compatibility and incompatibility. Many of the Life Elements compatibility and incompatibility is a straightforward determination based on whether or not they are found in Nature or if they are combined by man for useful purposes. However, the compatibility/incompatibility determination of other Life Elements is far more complex. Multiple factors are weighed in order to make such determinations. For example, though some Life Elements may be found together in Nature, they may also have toxic or even explosive reactions when combined directly. Therefore, there can be situations where these two Life Elements can get along well, but there will also be those that experience overall negative relations. From this example you can see the complexity of determining the compatibility/incompatibility of certain Life Elements. Much research has been conducted and is ongoing to improve the accuracy for each Life Element. Elementology™ is in its infant stage. Like any science it is not exact, there is still much to learn, and Elementology™ will evolve and improve over time.

It is a pretty neat concept and cool to think about what Life Element you are, and to then compare it to what your significant others' Life Element is, and what both of you might combine to form. Or, what each of your parents is and what they may combine to form; or what you and your best friends are and so on. For example, a couple very well known to me are Hydrogen (1) and Fluorine (9). These Life Elements are highly attracted to one another, however, they can react violently, and form many dangerous and sometimes deadly compounds. Thus, one may conclude that though they are highly attracted they may not be the best fit in the long-term. Unfortunately, this held true in this case as their marriage ended in a messy divorce. On a more positive note, another example of a couple well known to me is Oxygen (8) and Hydrogen (1). These two elements combine to make water which is essential to life. Oxygen (8) and Hydrogen (1) are both highly attracted to one another and form relatively stable and very useful and essential compounds. This couple has been happily together for 20 years.

As mentioned in a previous section of this book, it would be reckless to say that "who we are" is limited to our inborn qualities only. The world we live in and all that we have been exposed to will also shape who we are, how we react, and who we do or do not choose to interact with. Humans also have the amazing ability of the mind to enact free will. Though we have certain innate qualities we do not have to be bound to them. We have the ability to both recognize and improve upon our shortcomings. However, this can take considerable conscious effort and high intention to overcome. For example, if your Life Element is not compatible with a certain other Life Element, this does not mean you need to avoid them altogether. We as humans have the ability to execute high intention to overcome anything; including even the elements. If you are less likely to be compatible with another Life Element based on Elementology™ it does not mean all is lost. Instead, if you are aware of your learned tendencies and the characteristics of those you interact with; and you are also aware of and in tune with your inborn qualities and those of who you interact with, you can manage the strengths and weaknesses of one another in a manner that provides the most positive outcomes for both of you.

LIFE ELEMENT PROFILES

In our beginning, and in our end, we are nothing more than a compilation of elements. All of that we are, all that we need to survive, and all the things that we possess are elements. Given that we are completely comprised of elements, there is an undeniable connection between us and the elements - they are who and what we are. We are all elements

1 THE DAY OF HYDROGEN

LIFE ELEMENT HYDROGEN (H) - 1ST DAY BIRTH ELEMENT

CLASSIFICATION

HYDROGEN is a gas at standard temperature and pressure and thus possesses Air Element qualities. It is also one of the six "basis of life" elements - CHONPS (Carbon, Hydrogen, Oxygen, Nitrogen, Phosphorous, and Sulfur). It is also classified as a non-metal which means it is typically not a good conductor of heat or electricity. However, in reality, HYDROGEN should be in a class of its own as it is the only element that does not contain any neutrons. This provides it with unique characteristics aside from all other Life Elements.

COMPATIBILITY

Hydrogen personalities are most compatible with Life Elements Helium (2), Carbon (6), Nitrogen (7) and Oxygen (8). They are also very compatible with fellow HYDROGENs. Beryllium (4), Boron (5), Neon (10), Sodium (11), Magnesium (12), Phosphorus (15), Sulfur (16), Chlorine (17), Argon (18), Potassium (19), Vanadium (23), Chromium (24), Manganese (25), Iron (26), Nickel (28), and Copper (29) also get along well with HYDROGENs.

INCOMPATIBILITY

HYDROGENs are less likely to be suitable with Lithium (3), Aluminum (13), Silicon (14), Calcium (20), Scandium (21), Titanium (22), Cobalt (27), Zinc (30), and Gallium (31) personalities. HYDROGEN personalities can have tumultuous bonds with certain other Life Elements that they are highly attracted to. In particular, HYDROGEN personalities can have very passionate love-hate associations with Fluorine (9) which they are probably better off avoiding altogether. Also, though they can under certain circumstance get along very well, Chlorines (17) and HYDROGENs can also have rocky relations.

ELEMENT BACKGROUND, GENERAL DISPOSITION QUALITIES

HYDROGEN is the most common of all known elements in the Universe. Incredibly it accounts for 90% of all atoms by occurrence and 75% by weight. Along with Helium (2), HYDROGEN makes up 99.9% of the known elemental matter in the Universe. On Earth HYDROGEN ranks 3rd by occurrence on the surface but it is much lower in abundance by weight. This is because it is the smallest and simplest and thus the lightest of all elements. It is the 3rd most common Life Element in the human body. It likely has more unique compounds than any other Life Element on Earth. Even more so than Carbon (6) which has a mind boggling amount of different compounds as well. HYDROGEN, along with Carbon (6) form most of these compounds with each other. HYDROGEN was one of two of the very first elements created at the onset of the Big Bang. It is essential for all life as we know it. Despite its very low mass, HYDROGEN accounts for nearly 10% by weight of the elements that make up living beings. HYDROGEN, though very stable under normal conditions, is quite reactive at elevated temperatures and combines with almost all other elements. It is the building block of all other elements; HYDROGEN atoms under the right circumstances (typically in stars) combine through fusion to create other elements. That is, two HYDROGEN atoms can fuse together to form Helium (2) (nuclear fusion is different from elements forming compounds) and then build upon that to form other elements. HYDROGEN is one of the spiritual Life Elements, and is therefore closely connected with Nature. HYDROGENs are generally very social, get along well with others more so than most, and can have a knack for finding quick common grounds upon meeting many differing types of people. They are near the very top, if not at the top, when it comes to their ability to relate to others in some way. They also tend to be non-complex individuals with less drama than most. They are generally efficient, organized, and prefer to keep things simple; they may often be minimalists. Though they are simple and are typically non-complex, they are excellent at navigating and resolving complicated and delicate relations that can involve multiple parties. They tend to be well balanced with utilizing their emotions and intuition along with thought and analysis. Though generally quite stable, Hydrogens do not necessarily hold up well when cornered or pressured, or are highly stressed, or when they find themselves in the midst of a lot of drama. However, they can be a good source for calming and cooling down heated situations or individuals who are distressed. They also can be level headed, and tend to shine and take on leadership roles in dire situations. Many will be drawn to HYDROGENs in heated, contentious, and desperate situations; this attraction will likely linger beyond these dire situations or events. HYDROGENs are generally often found to be reasonable and practical persons.

ADDITIONAL DEFINING QUALITIES, CHARACTERISTICS & INTERESTS

HYDROGEN is colorless, odorless, and tasteless. It is flammable and burns with Oxygen (8) to form water. It has the lowest density of all gasses. HYDROGEN is a necessity for all animals, plants, and other terrestrial life. HYDROGEN is found free in Nature on Earth in very low quantities in its diatomic state (two bonded HYDROGEN atoms); it combines with itself. HYDROGEN personalities can be very open and transparent. This is a quality that is a major component of their appeal. They also tend to be very good observers; are more laid back and not "rock the boat" types; they can be quiet types too and may hold things close to their vest, but they can also at the same time be the life of a party. Although generally well-tempered, they can also have a fiery side to them as well. HYDROGEN personalities can often be content with who they are and their position in life. They also have a need for and enjoy their alone time. At the same time however, HYDROGEN personalities have an above average need for companionship. HYDROGENs enjoy interacting and particularly enjoy the give and take exchanges with others. HYDROGEN personalities have many interests and tend to be curious about life; they enjoy history, science, nature, politics, libations, entertaining, learning, home projects, the water, the Sun, the environment, plants/flowers/trees, animals, pets, and reading. They also tend to enjoy being active in general.

2 THE DAY OF HELIUM

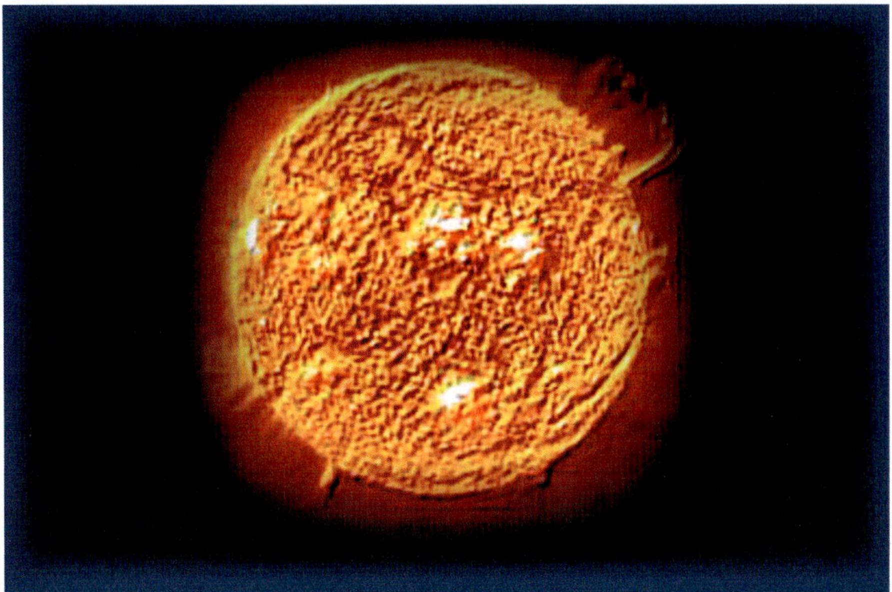

LIFE ELEMENT HELIUM (He) - 2ND DAY BIRTH ELEMENT

CLASSIFICATION

HELIUM is a gas at standard temperature and pressure and thus possesses Air Element qualities. It falls into the category of Noble Gases. As a Noble Gas, its outer ring is filled to the limit with electrons. Therefore, it is very stable and does not readily form compounds.

COMPATIBILITY

HELIUM personalities are most compatible with Hydrogen (1) and other HELIUMs. They can also relate well with Nitrogen (7), Oxygen (8), Neon (10), Sodium (11), and Argon (18). Additionally, Iron (26) can be very influential on HELIUMs.

INCOMPATIBILITY

HELIUM does not react directly with many, if any other Life Elements. They are less likely to have commonalities with Lithium (3), Beryllium (4), Boron (5), Carbon (6), Fluorine (9), Magnesium (12), Aluminum (13), Silicon (14), Phosphorus (15), Sulfur (16), Chlorine (17), Potassium (19), Calcium (20), Scandium (21), Titanium (22), Vanadium (23), Chromium (24), Manganese (25), Cobalt (27), Nickel (28), Copper (29), Zinc (30), and Gallium (31) versus other Life Elements.

ELEMENT BACKGROUND, GENERAL DISPOSITION QUALITIES

HELIUM is the second most common element in the Universe of known matter but interestingly it is quite rare on Earth. It makes up 9.9% of the remaining 10% of elements that exist outside of Hydrogen (1) which makes up 90% of all known elements in the Universe. Large amounts of HELIUM are continually created by nuclear fusion of Hydrogen in stars [the nucleus of two Hydrogen (1) atoms combine to form HELIUM]. HELIUM is the second smallest and simplest of all elements. HELIUM, along with Hydrogen (1), was among the very first elements created at the onset of the Big Bang. HELIUM is very stable and it does not react with any other element under standard conditions. HELIUM is not found in any living beings that we are aware of, but it is in every star. HELIUMs are the building blocks of many elements; HELIUM atoms fuse (two HELIUM atoms nuclei combine - this is different from elements forming compounds) in very large stars. HELIUMs will have more challenges relating to other Life Elements in comparison to others. They can sometimes find it difficult to befriend and relate to others in a deeper and more meaningful way. Typically, making friendships with strangers does not come easily to them. They also tend to be more guarded than most. They are simple individuals in the sense that they are easily content and satisfied, but at the same time can be rather complex in their minds and their inward thinking. In particular, they can have a fascination with existence, Nature, spirituality, the meaning of life, their purpose, complex issues/problems, etc. They are thinkers and often may think too much about certain issues for their own good. They innately favor thought and analysis over emotion and intuition in their assessment of others and situations in general. Like anyone, should they stray from their innate tendencies and in their case rely more upon emotion & intuition (as opposed to thought & analysis), their choices will more often not produce the best results. They tend to be non-judgmental, don't necessarily like to take sides, and are not easily swayed. As a result of their outward stability, non-judgmental qualities, and their calm demeanor, HELIUMs can carry on rather well with people despite their inward difficulties relating to others. They are also generally excellent at processing and managing stress and confrontation. They are likely not going to be ones who will stir up trouble. They are "go with the flow" type personalities. They can be well suited to defuse confrontation between others, and make great listeners, but they may not necessarily be the best at expressing their own emotions and feelings. They can come across as being somewhat "unreactive". Some may even come across as cold, but they are more often actually warm, loving, even-keeled, and tranquil people. HELIUMs can also come across as aloof or scatter brained at times but there is often much more to them than one might perceive. One should not underestimate them in this regard. They have no place for conflict and drama in their lives but can manage through it well from others as they are often quite tolerant individuals.

ADDITIONAL DEFINING QUALITIES, CHARACTERISTICS & INTERESTS

HELIUM is colorless, odorless and tasteless. HELIUM is not a necessity for terrestrial life. However, the Sun and all other stars are made of HELIUM. HELIUM has a very low density; and possesses the lowest melting and boiling points of all elements. HELIUM is a finite resource on Earth. HELIUM people can be complex in the sense that they are transparent but it may not be obvious to others. That is, they do not readily project their feelings as they tend to be less expressive than most. So, even though they are an open book once you get below the surface, it may take some extra initiative by others to see this side of them. They can be content being with themselves more so than most other people; they very much value their alone time. They get along great with other HELIUMs. They also tend to be more inward with their emotions and keep things to themselves. Like everyone, HELIUMs require some kind of companionship but they will often find themselves struggling to find those that are on their level. This can lead to HELIUMs settling more often in their friendships and relationships in comparison. Unfortunately, this quality may give rise to the odds that they will interact with others that they perhaps were better off avoiding. HELIUMs, when self-aware of their innate qualities, can learn to take extra caution to avoid certain individuals (and situations) when deciding who they interact with and befriend. HELIUMs also tend to be more laidback and easy going types; they also tend to be more reserved and quiet. They can also have an appreciation for aesthetics and be rather detailed; can be artistic, and also pleasing to the eye. HELIUM personalities tend to like nature, travel, animals/pets, space, stars, the water, and the Sun. Did I mention the Sun?! HELIUMs typically have an affinity for the Sun more than any other Life Element. Thus they also tend to enjoy just about anything that has them outdoors.

3 THE DAY OF LITHIUM

LIFE ELEMENT LITHIUM (Li) - 3RD DAY BIRTH ELEMENT

CLASSIFICATION

LITHIUM is a solid metal but is also flammable at standard conditions and thus possesses both Earth and Fire Element qualities. It falls into the category of Alkali Metals which means it typically gives up its only outer electron to form bonds with non-metals. It is malleable, ductile and an excellent conductor of heat and electricity. It is softer than most metals.

COMPATIBILITY

LITHIUM personalities are most compatible with and tend to get along with Boron (5), Carbon (6), and Oxygen (8), Sodium (11), Aluminum (13), Silicon (14), Phosphorus (15), Potassium (19), Titanium (22), Manganese (25), and Iron (26). LITHIUMs are also very attracted to Fluorine (9) and Chlorine (17) but they may not always be best long-term as these relations can sometimes be volatile.

INCOMPATIBILITY

LITHIUM personalities are likely to be less compatible with Hydrogen (1), Helium (2), Beryllium (4), Neon (10), Magnesium (12), Sulfur (16), Argon (18), Calcium (20), Scandium (21), Vanadium (23), Chromium (24), Cobalt (27), Nickel (28), Copper (29), Zinc (30), and Gallium (31). Though LITHIUM and Nitrogen (7) have an attraction and commonalities, they can also have many tensions between them. LITHIUMs are also very attracted to Fluorine (9) and Chlorine (17), but they may not always be best long-term as these relations can sometimes be volatile.

ELEMENT BACKGROUND, GENERAL DISPOSITION QUALITIES

LITHIUM is not one of the more abundant elements in the Universe or on Earth. However, LITHIUM was the third element created within the first few minutes of the Big Bang (Hydrogen and Helium being the first two). It is found in humans in trace amounts. LITHIUM is unique as it is the lightest metal and the least dense solid element. LITHIUM is corrosive, reactive, and flammable under normal conditions. LITHIUMs are innately spiritual; their spirituality can be expressed in varying ways. LITHIUMs often relate to the world from a different perspective than most other Life Elements; they can be viewed as "eclectics". They are generally pretty open, engaging and giving people. Some LITHIUMs can often show their excitable and fiery side which is not always appreciated by everyone. They can also unfortunately, often be misunderstood and come across to the contrary of their intentions. LITHIUMs desire to fit in and may sometimes try too much to achieve this. They can be both a source of and engage in drama. They can be known to be spirited and intense individuals too. LITHIUMs tend to be simple people, not necessarily in thought and mind but rather in their desires and needs. They appreciate simple things. Thoughts and intentions very often will mean much more to them than anything else. Thoughtful and well-intended actions by others towards them will be very much appreciated and will go a long way for those perpetrating the good deeds. LITHIUMs can also be very smooth and cunning in character when they want or need to be. They are innately governed far more by emotion and intuition over thought and analysis. They are more likely at their best when relying on their instincts rather than over thinking or over analyzing circumstances or people. LITHIUMs handle day to day stress moderately but are not very good with handling elevated and long-term stressors. LITHIUMs are sometimes known to be "old souls".

ADDITIONAL DEFINING QUALITIES, CHARACTERISTICS & INTERESTS

LITHIUM is a soft silvery metal that is malleable, ductile and an excellent conductor of heat and electricity. It easily reacts with water and quickly forms a black tarnish when exposed to air. It is found in trace amounts in all living organisms. It is believed to play an important role in the health of living beings and appears to have a link with extending human lifespan. However, it is unknown exactly what role LITHIUM plays in physiology. LITHIUM personalities require human interaction - they tend to not like to be alone, more so than most. They also tend to be more outward with their emotions and can find it difficult to keep things to themselves. LITHIUMs outgoing personalities and high need for human interaction can sometimes lead them to find themselves in friendships or relationships just for the sake of it. Rather than focusing and placing emphasis on the quality of those they interact with and the purpose of a relationship itself, they may overlook these values to satisfy their desire for companionship. However, when LITHIUMs are mindful of these qualities and are at peace with themselves, they can be quite introspective and self-aware. These qualities enable them to strike balance between their desires to be with others, how they come across to others, and what qualities others bring to them. When LITHIUMs are in tune with themselves, they will be far more likely to avoid the pitfalls of interacting with people that may not be the best fit for them.

LITHIUM people can be both very tough yet have a soft caring side; are versatile and can adapt well; can perform and sustain their cool in dire straits; are often energetic; and they can also be very "thick skinned" - they can both dish it out and take it. They are generally very open to forming new relationships and romances, perhaps too open at times. They also more often have very passionate relationships and even more so when paired with the right Life Element. LITHIUMs tend to appreciate aesthetics, the arts, medicine, science, religion/ spirituality, Nature, the water; and seek to maintain a healthy life/work balance.

4 THE DAY OF BERYLLIUM

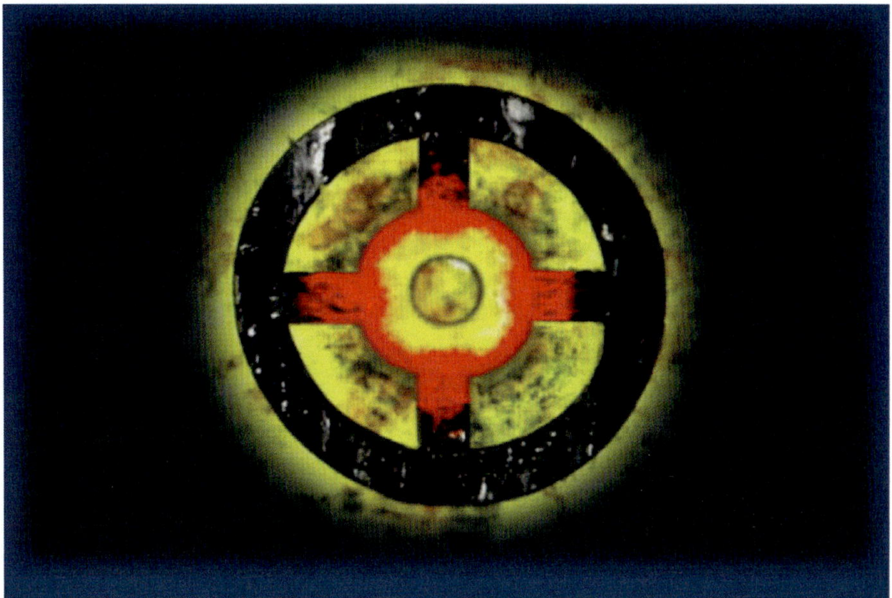

LIFE ELEMENT BERYLLIUM (Be) - 4TH DAY BIRTH ELEMENT

CLASSIFICATION

BERYLLIUM is an Alkali Earth Metal that is solid at standard temperature and pressure and thus possesses Earth Element qualities. It readily gives up its two outer electrons when it bonds. It is somewhat reactive at standard temperature and pressure. It is a good conductor of heat and electricity. It has low density; high melting and boiling points.

COMPATIBILITY

BERYLLIUM personalities are compatible with Oxygen (8), Aluminum (13), and Silicon (14). They also tend to get along well with Hydrogen (1), Boron (5), Fluorine (9), Sodium (11), Phosphorus (15), Chlorine (17) and Scandium (21). They can also relate with fellow Alkali Earth Metals Magnesium (12) and Calcium (20). Additionally, they also can form very mutually positive relations with Iron (26), Nickel (28), and Copper (29).

INCOMPATIBILITY

BERYLLIUM personalities are less likely to be compatible with Helium (2), Lithium (3), Carbon (6), Nitrogen (7), Neon (10), Sulfur (16), Argon (18), Potassium (19), Titanium (22), Vanadium (23), Chromium (24), Manganese (25), Cobalt (27), Zinc (30), and Gallium (31).

ELEMENT BACKGROUND, GENERAL DISPOSITION QUALITIES

BERYLLIUM is at the low end of abundance in relation to other elements in the Universe and on Earth. It is a trace element in the human body, it is barely detectable. Although there are many known compounds on Earth that contain BERYLLIUM, most are not very abundant. BERYLLIUMs can find it moderately easy to get along with many but may just as equally have some difficulty relating to certain other Life Elements. However, the difficulties they may have relating to others may be more often of their own perception and cause versus reality at times. BERYLLIUMs can over think their interactions and relations with others and second guess themselves. They may be better served by following their intuition along with a more measured amount of thought and analysis. BERYLLIUMs are intrinsically engaging, giving and charitable people. They often enjoy helping others in any way they can. They tend to be more outward with their emotions and to not keep things to themselves.

BERYLLIUMs also tend to require the company of others more than most. They require human interaction on a regular basis - they do not like to be alone. BERYLLIUMs engaging personalities and high need for human interaction does not as often lead to them settling in their friendships and relationships as one might think. That is because BERYLLIUMs tend to be a little more guarded and cautious with whom they allow close to them and who they romantically bond with. Typically, the bonds they do form with others are very strong and deep. The relations with those that they connect with also tend to be meaningful and stable. They are innately governed more by emotion and intuition over thought and analysis. In general, they are better than most when it comes to handling and dealing with stress and stressful situations.

ADDITIONAL DEFINING QUALITIES, CHARACTERISTICS & INTERESTS

BERYLLIUM is a hard, rigid, steel-gray metal that is an excellent conductor of heat and electricity. It is one of the lightest metals yet it has exceptional stiffness; it has the best heat dissipation of all metals per unit weight; it also has a very high modulus of elasticity - 50% greater than that of steel. BERYLLIUM is very resistant to wear, even at very high temperatures. BERYLLIUM has no biological role and is toxic. BERYLLIUM personalities can be stern yet come across easy tempered; straightforward and strong-willed. They can be very good mediators, especially in highly confrontational situations. They also tend to be dedicated, focused, and unwavering when there is a goal to be achieved. BERYLLIUMs tend to be the "tough" in the expression "...when the tough get going". BERYLLIUMs tend to be good communicators and trustworthy with keeping things to themselves. They prefer to view things from an aspect that allows them to see people and situations beyond the surface. That is, they may withhold judgment until they are presented with a full "picture". One result of this is that they may struggle with their work, or certain other problems and issues if they do not have enough detailed and in-depth information to make a solid conclusion. BERYLLIUMs generally tend to be more interested in things outside of Nature but they are people oriented. They have an affinity for jewelry, "bling", and cars; they also tend to be active, like sweets, music, art, medicine, science, aeronautics, space/space exploration, history, and travel. They tend to have a stronger affiliation for their nationality in relation to other Life Elements; and also may have interests or affiliations with the military more so than most.

5 THE DAY OF BORON

LIFE ELEMENT BORON (B) - 5TH DAY BIRTH ELEMENT

CLASSIFICATION

BORON is a metalloid and exists as a solid at standard temperature and pressure and thus possesses Earth Element qualities. However, being the metalloid that it is, BORON also possesses both metal and non-metal characteristics. It is one of only two Life Elements that are metalloids, Silicon (14) being the other. It typically shares its outer electrons to form stable bonds.

COMPATIBILITY

BORON personalities are most compatible with Oxygen (8), Sodium (11), and Calcium (20). They can also relate with Hydrogen (1), Lithium (3), Beryllium (4), Nitrogen (7), Fluorine (9), Magnesium (12), Silicon (14), Chlorine (17), Potassium (19), Titanium (22), and Manganese (25).

INCOMPATIBILITY

BORON personalities are less likely to be compatible with Helium (2), Carbon (6), Neon (10), Aluminum (13), Phosphorus (15), Sulfur (16), Argon (18), Scandium (21), Vanadium (23), Chromium (24), Iron (26), Cobalt (27), Nickel (28), Copper (29), Zinc (30), and Gallium (31).

ELEMENT BACKGROUND, GENERAL DISPOSITION QUALITIES

BORON is somewhat rare in abundance in relation to other elements in the Universe but it is toward the middle on Earth. It ranks 23rd in occurrence in the human body. Most BORON exists in two forms. It is reactive in its amorphous form, but un-reactive (inert) in its crystalline form. It is not surprising that it is the second hardest element known (in its crystalline form) given that it is neighbors with Life Element Carbon (6) which is the hardest element (in diamond form). At room temperature it is an insulator but at high temperatures it is a conductor. These varying qualities are characteristic of metalloids which make them somewhat unpredictable. Recent evidence suggests that BORON is essential for bone health in all animals and healthy brain functions such as memory and hand-eye coordination. BORON is also essential for plant life as it plays a crucial role in the cell walls of plants. Plants that lack BORON turn yellow, wither and die. People born on the 5th day will find it easy to relate with certain Life Elements. However, they will also find it difficult to relate to many others as well. BORONs can be outgoing but at the same time can be just as easily introverted and come across as stand-offish and rigid. They are often complicated and contradictory. BORONs enjoy interaction and the company of others, but this may only be true in many cases so long as the company and interaction are of their choosing. They can be methodical, analytical, and detailed. They will prefer to cut to the chase and call out issues, problems, or things that cause negativity or unproductiveness in any way. They will take the lead or be of great aid in the resolution of such matters. They are creative, curious and enjoy resolving problems. They may be viewed as "beat to the sound of their own drum" and also be "stick to their guns" type people; these are not necessarily meant to be interpreted as negative qualities. They can also be generally unpredictable and can often be very difficult to figure out, even by themselves. They make for good negotiators; can be great at resolving conflict; and can be known to keep their calm in very difficult and heated situations. They tend to handle both day to day and long-term stress better than most. They have a good balance with utilizing thought and analysis versus intuition and emotion when assessing situations and others. BORONs, despite having some unpredictable characteristics are typically quite reliable and stable individuals.

ADDITIONAL DEFINING QUALITIES, CHARACTERISTICS & INTERESTS

BORON is black-brown-silvery, very hard, brittle, and very resistant to heat in its crystalline form. It has the second highest melting point of all elements; second only to Carbon (6). BORON is not soluble in, nor does it react with water, air or acids at standard conditions. There is much to learn about its role in physiology, but we do know that it is essential for life even though it is only found in very trace amounts in living things. BORON has been found to be extremely attracted to cancer cells in the body (while ignoring normal cells) which has opened up extensive and promising research for cancer treatments. BORON personalities can be perceived as tough, straightforward and strong-willed. However, they can be some of the most accepting and understanding people; they can also be very supportive to others in everyday life as well as in hard times too. Many may be surprised to learn how caring they actually are of others as this may not be apparent and they may actually seem contrary to such. In other words, they can come across as having a hard shell but inside they are Teddy Bears. Thus, it should be no surprise that they are often not the most expressive with their emotions to most other people. However, for those who are closest to them, they have little reservations about expressing their inner thoughts and emotions. BORONs can also be one of the most sharing of all the Life Elements. They also tend to be sentimental, and as a result may hold onto some things longer than most people - these can be physical, mental or emotional. They tend to have interests in cooking, baking, crafts, music, electronics (computers, phones, audio, etc.), plants, flowers, nature, outdoors, animals, medicine, archeology, biology and playing games (brain teasers, electronic, board, etc.). BORONs can also have an affection for cleanliness, primping and caring about their appearance. They tend to have a disliking for insects (more than most).

6 THE DAY OF CARBON

LIFE ELEMENT CARBON (C) - 6TH DAY BIRTH ELEMENT

CLASSIFICATION

CARBON is a solid nonmetal at standard temperature and pressure and thus possesses Earth Element qualities. It is also one of the six "basis of life" elements - CHONPS (Carbon, Hydrogen, Oxygen, Nitrogen, Phosphorous, and Sulfur). It utilizes its four outer electrons to form bonds that share these electrons with another elements electrons. Though it is a nonmetal, CARBON is unique in that it can take on many different forms and thus possess varying properties.

COMPATIBILITY

CARBON personalities are most compatible with Hydrogen (1), fellow CARBONs, Nitrogen (7), Oxygen (8), Magnesium (12), Sulfur (16), and Calcium (20). They can also relate well with Lithium (3), Fluorine (9), Sodium (11), Aluminum (13), Chlorine (17), Potassium (19), Chromium (24), Manganese (25), Iron (26), Cobalt (27), Nickel (28), Copper (29), Zinc (30), and Gallium (31).

INCOMPATIBILITY

CARBON personalities are less likely to be compatible with Helium (2), Beryllium (4), Boron (5), Neon (10), Silicon (14), Argon (18), Scandium (21), Titanium (22), and Vanadium (23). Though CARBON and Nitrogen (7) form a vast array of essential biological compounds, they also form many toxic compounds and hence they don't always make for the best relations. Additionally, Phosphorous (15) and CARBONs though highly attracted to one another may not necessarily make for the best relations in the long run.

ELEMENT BACKGROUND, GENERAL DISPOSITION QUALITIES

CARBON is at the high end of abundance in relation to other elements in the Universe and on Earth. It is the 4th most common element in the Universe, 10th in the Earth's crust, and the 2nd most common in the human body. CARBON is a unique element in many ways. CARBON is the basis of life as we know it. CARBON has millions of known combinations with more continually being discovered. It can form very long stable chains with itself and other elements. CARBONs will tend to find it very easy to relate with most other Life Elements. They are also typically social and outgoing more than most. CARBONs can have very diverse interests, social circles and perspectives. They are extremely adaptable and respond well to change. They can vary and adapt so well that they can have chameleon like personalities. CARBONs can be extremely difficult to predict; and they can sometimes be fickle and indecisive. In certain situations, some CARBONs can be very good liaisons, leaders, in charge, understanding, calm, stern and collected. Though in other circumstances some can be just the opposite or even a combination of the two. Generally though, CARBONs tend to be energetic, curious, creative, rather complex and deep, and be excellent communicators. CARBONs will often make for good managers, leaders, thinkers/analysts, and find creative solutions. They appreciate complex issues and problems and can find very creative ways to resolve them. CARBONs can also be very engaging and affectionate. They enjoy sharing what they have with others and appreciate when others reciprocate those offerings. Compared to most others, their intentions are more altruistic opposed to sharing for personal or even mutual gain. They are most often capable of handling day to day and acute stressors better than almost all other Life Elements. CARBONs are more often content people; can be very happy by themselves and live long fruitful lives by themselves (from a romantic perspective). However, they love people and the company of others and therefore have a high need for the company of others. CARBONs are lovers of life and enjoy & appreciate all life forms more than any other Life Element. They are also somewhat unique in that they can utilize both sides of their assessment capabilities better than most. That is, they can rely on both their emotions and intuition and thought and analysis more so than any other Life Element. CARBON is truly a unique and incredible Life Element and often CARBON personalities are found to be just that.

ADDITIONAL DEFINING QUALITIES, CHARACTERISTICS & INTERESTS

CARBON is essential to all life on Earth as we know it, including both plants and animals. A major reason for CARBON being the basis of life is that it forms an enormous amount of structured chains of differing and very long lengths. In fact, there are more known CARBON compounds than all other elemental compounds combined other than HYDROGEN (1) which is involved with most CARBON compounds. It also has an incredible ability to bond with many other elements (some in standard conditions and others in non-standard conditions). In fact, CARBON will combine with all of the other Life Elements in the right conditions except Helium (2) and Neon (10). CARBON is resistive to both acids and corrosion; and has the highest melting point of all the Life Elements. CARBONs tend to "share" their thoughts and emotions with others rather than engage in lop-sided relations. That is, they seem to prefer to exchange their feelings and emotions with those that they interact with. They will feel uneasy having relations with those where one person dominates over the other. They don't mind listening but they are more drawn to people that they can have an interactive exchange of ideas, emotions, thoughts, and feelings. They have a pretty healthy expression of their emotions. CARBONs love interaction, whether it be with people, animals or nature. They prefer and tend to excel in structured environments. They may be regimented, neat, organized, and structured in their own lives both personally and professionally. They can also be very detailed individuals. CARBONs are gifted in the way that they are capable of getting along with so many different personality types.

CARBON exists in multiple forms such as graphite, diamond, and amorphous (shapeless, powder-like) states. It is stable in some forms (graphite & diamond) in nature and not so much in other forms (amorphous). Its electrical and thermal properties vary to the extreme. For example, graphite is one of the softest substances, while diamond is the hardest naturally-occurring material known; graphite is a very good electrical conductor while diamond has very low electrical conductivity; and diamonds have the highest thermal conductivity of all known materials. CARBON is very widely found throughout nature and in man-made items. CARBONs can have a wide range of interests and will often enjoy taking on new hobbies & interests with others. They love Nature, the environment, and people. Some other interests of theirs include pets, animals, water, the Sun, most anything outdoors, observing and playing many types of sports and recreation, entertaining, food, cooking, libations, travel, clothes, writing, reading, technology, the arts, jewelry, clubs/groups, being active in their community, interior/home décor, and exploration.

7 THE DAY OF NITROGEN

LIFE ELEMENT NITROGEN (N) - 7TH DAY BIRTH ELEMENT

CLASSIFICATION

NITROGEN is a gas at standard temperature and pressure and thus possesses Air Element qualities. It is also one of the six "basis of life" elements: CHONPS (Carbon, Hydrogen, Oxygen, Nitrogen, Phosphorous, and Sulfur). NITROGEN has 5 electrons in its outer shell and it can form three bonds with three different elements or less at the same time - this is uncommon.

COMPATIBILITY

NITROGEN personalities are compatible with and tend to get along best with other NITROGENs. They can also get along quite well with Hydrogen (1) and Oxygen (8). They also relate with Carbon (6), Fluorine (9), Sodium (11), Sulfur (16), Chlorine (17), Potassium (19), Calcium (20), Titanium (22), and Nickel (28); and to a lesser degree they relate with Helium (2), Boron (5), Neon (10), Phosphorus (15), and Argon (18).

INCOMPATIBILITY

NITROGEN personalities are less likely to be compatible with Beryllium (4), Aluminum (13), Silicon (14), Scandium (21), Vanadium (23), Chromium (24), Manganese (25), Iron (26), Cobalt (27), Copper (29), Zinc (30), and Gallium (31). NITROGEN and Carbon (6) can relate very well with one another and share a strong mutual attraction. Unfortunately, NITROGEN and Carbon (6) form many explosive and toxic compounds and thus may not necessarily always make for the best relations. Additionally, although NITROGEN has an attraction and commonalities with both Lithium (3) and Magnesium (12), they can also have some tensions with these two elements. Similarly, although NITROGENs and Titaniums (22) are attracted to one another, they may be better off not engaging in relations.

ELEMENT BACKGROUND, GENERAL DISPOSITION QUALITIES

NITROGEN is at the high end of abundance in relation to other elements in the Universe and on Earth. It is the 7th most common Life Element in the Universe. It makes up 78% of the air we breathe and approximately 3% by mass of the human body. However, NITROGEN is not very reactive under normal conditions. It typically forms very stable compounds especially with itself. It is the most abundant uncombined Life Element on Earth (combined with itself). Oddly, despite being the most abundant uncombined element, it is the 4th most electronegative Life Element (measurement of how much it attracts other elements electrons). With such a high electronegativity it would seem that it would be one of the least uncombined rather than the highest. NITROGENs are pretty equally balanced when it comes to relating with other Life Elements. That is, they get along great with just about the same amount of Life Elements that they do not necessarily have much in common with. They can be alluring and interesting to many, even those that they may not necessarily have innate commonalities or attractions with. They thrive on interaction with others on both a friendship and romantic level. Nonetheless, they can often keep their "distance", and withhold many of their innermost thoughts and feelings from others - not necessarily a bad thing in many cases. As a result of keeping their "distance", they do not necessarily form quick, deep/meaningful bonds with others. That is, NITROGENs can be very open, friendly and very easy to get along with, but they are innately reserved and perhaps leery when it comes to forming deep friendships and relationships. They by no means hold back from getting to know others, but they tend to take their time letting others into their inner circle. Though the relationships they form with others may come slowly, once they allow others "in", they will form some of the strongest and deepest bonds. They may often be viewed as "social butterflies". They may struggle internally romantically because of their strong desires to be with others versus their often "cautious" approach. NITROGENs care for those close to them very deeply and typically love with intensity and passion. They are also very loyal to those that are close to them. They are governed more by thought and analysis over emotion and intuition. They are typically not a source of drama but can tolerate it pretty well. They are also capable of handling stress very well, in fact they process and handle it better than most. They are also typically reliable, stable, trustworthy, cautious, can have a calming presence, and are not easily swayed by others. They can often have solid advice to offer and be "the voice of reason" in many situations.

ADDITIONAL DEFINING QUALITIES, CHARACTERISTICS & INTERESTS

NITROGEN is colorless, odorless and tasteless. NITROGEN can combine with itself alone and is found on Earth mostly in its diatomic state (two NITROGEN atoms combined together). NITROGEN is essential for all life as we know it. However, without two certain types of bacteria in the soil we would not exist because plants and animals are incapable of converting NITROGEN into a form that is useful. These certain kinds of bacteria found in the soil invade the roots of plants and convert NITROGEN into a useful form in plants. The plants are then consumed by animals to provide us with the necessary NITROGEN we need. NITROGEN does not readily react with acids, bases, oxidants, or reductants. NITROGEN personalities can be very content with who they are; get along very well with other NITROGENs; and can very much enjoy time to themselves. They are generally very stable individuals. They can also be inward with their thoughts and emotions while still being very open with others. NITROGENs are more selective than most with who they open up to. Once they do open up they can be very protective and sometimes become dependent and clingy with those individuals. Also, under certain conditions and with particular individuals NITROGENs can have explosive interactions. NITROGEN people can generally be open and transparent. However, this does not translate into every facet of their lives. For example, they are not necessarily as open and transparent as Oxygen (8) personalities might be to nearly everyone in nearly all situations. This is because of NITROGENs innate characteristic to be cautious with whom they allow into their lives on a deeper level. However, once allowed into their "inner circle" they are atop the most open and transparent personalities. NITROGENs generally tend to have an affinity for life, animals, plants, the outdoors, and nature in general. NITROGENs tend to be comfortable in their own skin and are not easily swayed. This is not to say they are stubborn, as they are typically open to others views and to change as well. In fact, they appreciate honest feedback and advice from others more so than most. They also tend to have a fondness for recreation, gardening, electronics, clothes, cleanliness, cooking, flowers/plants/trees, the wilderness, pets, travel, medicine, science, the water, and fashion. They also tend to enjoy "getting out and about" and being in groups, social events (concerts, shows, etc.), clubs, etc.

8 THE DAY OF OXYGEN

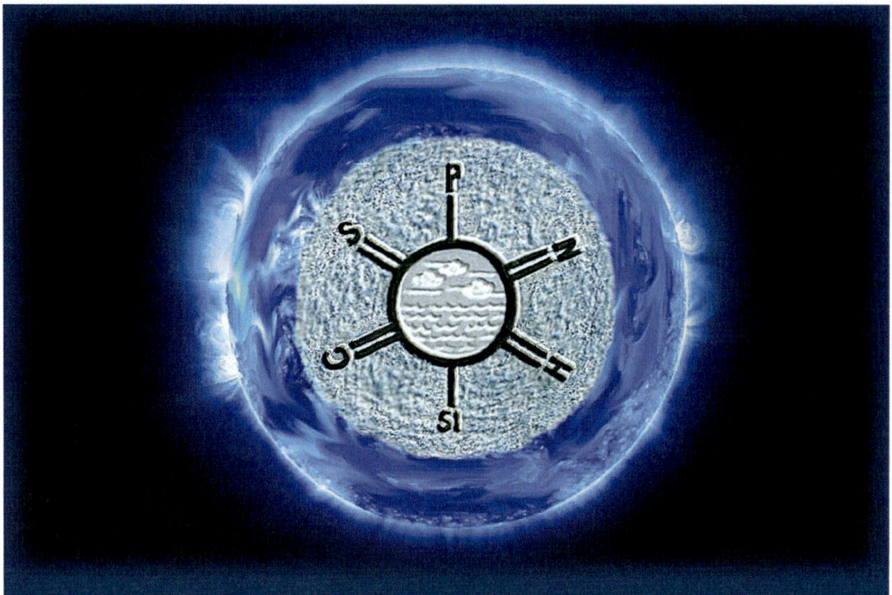

LIFE ELEMENT OXYGEN (O) - 8TH DAY BIRTH ELEMENT

CLASSIFICATION

OXYGEN is a gas at standard temperature and pressure and thus possesses Air Element qualities. It is one of the six "basis of life" elements - CHONPS (Carbon, Hydrogen, Oxygen, Nitrogen, Phosphorous, and Sulfur). It is classified as a non-metal which means it is not a good conductor of heat or electricity. It will almost always take 2 electrons when forming bonds with other elements and compounds.

COMPATIBILITY

OXYGENs tend to be very attracted to and get along best with Silicon (14). They are also very compatible with other fellow OXYGENs, Hydrogen (1), Lithium (3), Beryllium (4), Boron (5), Carbon (6), Nitrogen (7), Magnesium (12), Aluminum (13), Phosphorous (15), Sulfur (16), Potassium (19), Calcium (20), Manganese (25), Cobalt (27), and Gallium (31). Additionally, they can also relate with Helium (2), Neon (10), Argon (18), Nickel (28), Copper (29), and Zinc (30). Though OXYGENs have an attraction and commonalities with Sodium (11), Scandium (21), Titanium (22), Vanadium (23), Chromium (24), and Iron (26), these Life Element combinations can have tumultuous relations.

INCOMPATIBILITY

OXYGEN is highly reactive and is an oxidizing agent (takes electrons from other elements). OXYGENs may not be the best fits with Fluorine (9), and Chlorine (17). OXYGEN personalities, though easily forming friendships and relationships with nearly all others, they can have tumultuous bonds with certain other Life Elements that they have commonalities and attractions with. In particular, 8th day personalities can have turbulent associations with Sodium (11), Scandium (21), Titanium (22), Vanadium (23), Chromium (24), and Iron (26).

ELEMENT BACKGROUND, GENERAL DISPOSITION QUALITIES

OXYGEN is the 3rd most common element found in the Universe. OXYGEN is the most common element on the Earth's surface and in the human body. The oceans also consist of 90% OXYGEN. It has the 2nd highest electronegativity (an indicator of how much it attracts other elements electrons). It is essential to the survival of all aerobic life forms. OXYGEN will combine with nearly all other Life Elements. People born the 8th day tend to have more in common with others over any other Life Element. They typically will get along well with nearly everyone even if it pains them to do so. Should they need to, they can be some of the best at carrying on amicable shallow relations with those they may have disdain for. They are capable of easily establishing a quick rapport with others. They also generally can make friendships and form quick bonds with many different people from varying walks of life. They seem to have a most intuitive ability to find common ground with most other people. They can also be a calming presence to many and make others feel very comfortable in their presence. OXYGENs are not the best when it comes to dealing with stress. However, like everyone's, their stress can be managed. OXYGENs or any other Life Elements inability to recognize or appropriately process and handle stress will lead to disruptions to themselves and those around them. OXYGENs do not care much for drama and will consciously do all they can to avoid creating it. However, they can unintentionally be the cause of it. The drama they stir is often related to the unintended consequences of their mishandling of their own internal issues/stressors. They are typically very accepting of others and their lifestyles; and of situations, circumstances, and generally of the way things are. They are pragmatic and can be minimalists. They may often set a high standard for themselves with respect to their honor, values, and integrity. They can also be methodical and orderly in their approach to tasks, issues, relations, and situations. OXYGENs are innately governed more by emotion and intuition versus thought and analysis. However, this should not be interpreted that OXYGENs are not thinkers. In fact, the opposite is most often true as OXYGENs are typically very deep thinkers, introspective, and insightful. Despite that, they are best served by allowing their instincts to play the major role in their assessment of situations and others.

ADDITIONAL DEFINING QUALITIES, CHARACTERISTICS & INTERESTS

OXYGEN is colorless, odorless, and tasteless. It exists in Nature in its diatomic state (two OXYGEN atoms combined). OXYGEN is a necessity for nearly all terrestrial life. As mentioned, OXYGEN is capable of combining with itself alone. OXYGEN personalities can be very open and transparent. This furthers their appeal which contributes to their ability to find commonalities, attract, and bond with others. They also tend to have keen interpersonal assessment skills. They also relate very well to others of the same Life Element (OXYGEN). They can be very content being with themselves and very much enjoy their alone time. At the same time, OXYGEN personalities have a higher need for companionship. They can also be extremely protective and loyal to those that are close to them. They tend to be calm, gentle, caring souls that are easily satisfied. OXYGEN personalities tend to have a flair for or fascination with life; an enchantment with the physical world; a love of all living things including animals, birds, plants, flowers, etc. They also enjoy history, medicine, science, the military, nature, pets, outdoors, water (rivers, lakes, and oceans), the sun, the heavens, spirituality, travel, sports, staying active, and mechanics. They can also appreciate aesthetics especially in art, home décor, architecture, and construction. They also tend to have a general curiosity for how things work.

9 THE DAY FLUORINE

LIFE ELEMENT FLUORINE (F) - 9TH DAY BIRTH ELEMENT

CLASSIFICATION

FLUORINE is a gas at standard temperature and pressure and thus possesses Air Element qualities. FLUORINE is a Halogen and thus extremely reactive. FLUORINE is one of two Halogen Life Elements - Chlorine (17) being the other. It will typically take an electron from other elements or compounds when it bonds.

COMPATIBILITY

FLUORINE personalities are most compatible with Calcium (20). They also are very attracted to fellow FLUORINEs. They can also relate well with Life Elements Beryllium (4), Neon (10), Sodium (11), Magnesium (12), Aluminum (13), Phosphorus (15), Potassium (19) and Manganese (25). They can also get along with Lithium (3), Boron (5), Carbon (6), Nitrogen (7), Silicon (14), Sulfur (16), and Argon (18).

INCOMPATIBILITY

FLUORINEs are somewhat unique in that many of the Life Elements that they are less likely compatible with, they are actually highly attracted to. Some of these include Lithium (3), Oxygen (8), Chlorine (17), Iron (26), and Nickel (28). They also may not be the best fit with Helium (2), Scandium (21), Titanium (22), Vanadium (23), Chromium (24), Cobalt (27), Copper (29), Zinc (30), and Gallium (31). They can have very rocky relationships with certain other Life Elements despite having many commonalities and an attraction. In particular, 9th day personalities can have very passionate love-hate associations with Hydrogen (1). Additionally, though there is a mutual attraction, FLUORINEs might be better off avoiding Phosphorus' (15) altogether.

ELEMENT BACKGROUND, GENERAL DISPOSITION QUALITIES

FLUORINE is the 24th most abundant element in the Universe. The 13th most abundant on the Earth's surface and 14th in humans. FLUORINE is the most electronegative (attracts other elements electrons) element and is therefore the most reactive of all elements. As a result, it easily combines with most elements. However, many of its reactions are violent because of its extremely reactive nature. Astonishingly, there is no other chemical substance that can break FLUORINE from any of its naturally occurring compounds. FLUORINEs typically have little issue with approaching or engaging with others. It is common for FLUORINEs to be outgoing and gregarious. However, FLUORINEs can also exhibit contentious filled interactions with certain other Life Elements both in their usual day to day dealings and their personal relationships. FLUORINEs tend to have the strongest, deepest, and most passionate romantic relationships over any other Life Element. Their relationships can also be the most difficult to break off. They will also have deep, substantive, and long lasting friendships too. Typically those who are involved with FLUORINEs recognize and are accepting of their "unconventional" ways. FLUORINEs are very emotionally driven. FLUORINEs can be good listeners. They are very accepting and understanding of others so long as they approach others' situations without bias which they can sometimes do. FLUORINE personalities do not tend to keep things close to themselves. And they may sometimes do or say much more than they probably should. Their emotions may overrule their inhibitions. As a result their opinions are often heard more often than perhaps they should be. They can also be more likely than others to release their opinions and emotions in volatile fashions. Limiting tensions between both those who they come across in passing and those that are close to them should be a focus for them. FLUORINEs need to make extra conscious efforts to recognize and appropriately process and control their emotions and stressors so they can avoid distress, conflict, and breakdowns that they are innately prone to. FLUORINEs are emotional creatures that can hold onto to things (typically people, thoughts and emotions) more than any other Life Element. They will often have issues letting go of many things. They will also go to very extreme measures to keep others from interfering with their personal relations. They can be a source of drama and can also thrive on it. FLUORINEs typically handle day to day stress below the median level and will become unreasonable when they do not recognize acute or the long-term build-up of stressors and act accordingly. Given they have the highest electronegativity of all elements, FLUOERINEs need to make extra efforts to recognize and curb their emotionally and reactionary driven character so they can provide better balance in their lives. If they can achieve this balance they will exhibit the more favorable qualities of their personalities.

ADDITIONAL DEFINING QUALITIES, CHARACTERISTICS & INTERESTS

FLUORINE is a highly toxic pale yellow diatomic (it combines two FLUORINE atoms together) gas in its natural state. Thus, FLUORINE combines with itself alone. FLUORINE is very highly reactive and even explosive with certain other elements. It also reacts violently with water. Though it is still unknown exactly the entire role it plays, FLUORINE is most likely an essential trace element for animal teeth and bones. FLUORINE personalities can be content on their own, get along very well with other FLUORINEs, and very much enjoy their alone time. At the same time, FLUORINE personalities have one of the highest needs and desires for companionship. They can be very outward with their emotions. FLUORINE personalities can be prone to dependency issues; most typically dependencies on other people. It would behoove them to recognize and minimize any dependency tendencies as these can wreak havoc on them and severely alter their life path. Though some FLUORINE personalities can be very friendly and outgoing, they may have had, and may continue to have many struggles to achieve the outgoing and engaging people that they are. In other words, though they can get along well with many, this does not necessarily come easily to them. They tend to have had to or continually inwardly work harder than most in order to maintain the amiable and gregarious disposition portion of their personalities. Their interests tend to focus more on their relationships and people in their lives. They also tend to be neat, organized, energetic, clean, health conscious, and appreciate crafts and the arts. They also have interests in cooking, baking, clothing, movies/TV, and medicine. FLUORINE personalities have a very high need for companionship, intimacy, and are extremely loyal. However, once a friendship or relationship is broken, they are not easily repaired as they typically will not end well.

10 THE DAY OF NEON

LIFE ELEMENT NEON (Ne) - 10TH DAY BIRTH ELEMENT

CLASSIFICATION

NEON is a gas at standard temperature and pressure and thus possesses Air Element qualities. It falls into the category of Noble Gases. As a Noble Gas, its outer ring is filled to the limit with electrons. Therefore, it is very stable and does not readily form compounds with itself or any other element.

COMPATIBILITY

NEON personalities are most compatible with and tend to get along best with others most like them, namely, other NEONs. They can also relate with Hydrogen (1), Helium (2), Nitrogen (7), Oxygen (8), Fluorine (9), Sodium (11), and Argon (18).

INCOMPATIBILITY

NEON, like Helium (2) and Argon (18) do not react directly with many if any other Life Elements. Lithium (3), Beryllium (4), Boron (5), Carbon (6), Magnesium (12), Aluminum (13), Silicon (14), Phosphorus (15), Sulfur (16), Chlorine (17, Potassium (19), Calcium (20), Scandium (21), Titanium (22), Vanadium (23), Chromium (24), Manganese (25), Iron (26), Cobalt (27), Nickel (28), Copper (29), Zinc (30), and Gallium (31).

ELEMENT BACKGROUND, GENERAL DISPOSITION QUALITIES

NEON is the 5th most common element in the Universe but it is very rare on Earth. It is not found in humans or other life forms on earth as far as we know. NEON is continually created by the fusing of Helium and Oxygen in stars. NEON is found in stars in great abundance. On Earth, NEON is very stable and it does not react with any other element under normal conditions. Compared to most, people born the 10th day can have challenges relating with other Life Elements. They may find it more difficult to form many close meaningful relationships with other Life Elements. Though they can have very agreeable dispositions and have many friendships, the friendships are often not necessarily deep or long-lasting. This could be in part a result of their innate characteristic of being very content on their own. They are perhaps the most reclusive of all Life Elements. NEONs are not known to be the life of the party but they can be very social and easily fit in quite well with the crowd despite their lack of forming many close friendships with other Life Elements. NEONs can be good at handling combative and confrontational situations. They tend to be non-judgmental, don't necessarily take sides without overwhelming evidence, and are not easily swayed. They generally are calculated and ruled by thought and analysis versus emotion and intuition. They prefer to distance themselves far from drama. However, they are tolerant of it. They compartmentalize and process stress extremely well. However, when over pressured and pushed to their limits, which is rare, they can exhibit extreme reactions. They can offer unique perspectives for solving problems; and tend to have an ability to provide clarity to complex confusing matters. They are often level-headed, focused, successful, intelligent, goal oriented, straightforward, and "tell it like it is" people. They are typically not the best at expressing their inner emotions and feelings. However, outwardly they can often be known to be cool, calm, and collected people.

ADDITIONAL DEFINING QUALITIES, CHARACTERISTICS & INTERESTS

NEON is colorless, odorless, and tasteless. NEON is not necessary for any terrestrial life as far as we know. However, it is very common throughout the Universe. NEON has a very low density; and low melting and boiling points as well. It is not susceptible to breaking down to acids, bases; or oxidation/reduction. NEON people can be very open and transparent. As mentioned, NEONs are generally straightforward "tell it like it is" people. They tend to be more laid back, reserved, don't arouse easily, and can be protective of those they are close to. Some NEONs can be so no-nonsense and dry that they can come across as disassociated or disinterested; yet more often they are engaging in their own way. They are also often very loving and caring people to a surprising degree; their reserved nature that comes across can be misleading. NEON personalities tend to enjoy or require alone time more than most. They are also private people more so than most. They tend to be inward with their emotions and keep things to themselves. NEONs can spend extended periods of time happily on their own. NEONs like everyone to varying degrees, require companionship, but they can struggle more than most to find deep connections with others. This can lead to NEONs settling with who they have relations with. Unfortunately, this quality gives rise to the odds that they will interact with others that they perhaps were better off avoiding. NEONs who are in tune with their inner instincts will resist the temptations of settling or mingling with those that they should avoid. NEONs tend to appreciate a wide variety of aesthetics [natural occurrences (sunsets, colorful skies, exotic animals, beautiful landscapes and scenery, etc.), architecture, clothing, art, photography, how others look, just to name some] and can be concerned with their own appearance. They enjoy movies/TV, the theatre, dining, cooking, exercise, health conscious, nightlife, and the Sun. They also can be very spiritual/religious and may have a fascination with stars, the Universe and its "existence".

11 THE DAY OF SODIUM

LIFE ELEMENT SODIUM (Na) - 11TH DAY BIRTH ELEMENT

CLASSIFICATION

SODIUM is an Alkali Metal that is solid at standard temperature and pressure and thus possesses Earth Element qualities. SODIUM also possesses Fire Element characteristics as well because of its tendency to spontaneously ignite when exposed to water. It typically gives up its outer electron when forming bonds. It is malleable, ductile, and an excellent conductor of heat and electricity. It is softer than most metals and can be cut with a knife.

COMPATIBILITY

SODIUM personalities are most compatible with Chlorine (17). They also get along very well with Oxygen (8), Aluminum (13), Silicon (14), Potassium (19), and Calcium (20). They also relate to a lesser degree with Hydrogen (1), Helium (2), Lithium (3), Beryllium (4), Boron (5), Carbon (6), Nitrogen (7), Fluorine (9), Neon (10), Magnesium (12), Phosphorus (15), Sulfur (16), Titanium (22), Vanadium (23), Manganese (25), Iron (26), and Copper (29).

INCOMPATIBILITY

SODIUM personalities are less likely to be compatible with Helium (2), Argon (18), Scandium (21), Chromium (24), Cobalt (27), Nickel (28), Zinc (30), and Gallium (31). Though they attract and have commonalities with each other, SODIUMs and Oxygen (8) can have turbulent relations so they too may not always be the best fit.

ELEMENT BACKGROUND, GENERAL DISPOSITION QUALITIES

SODIUM is the 15th most abundant element in the Universe. On the Earth's surface it ranks as the 6th most abundant Life Element and in the human body it is the 9th most common. SODIUM is very reactive and thus reacts with a wide variety of other Life Elements. Sodium is an essential element to all animals and many plants. People born on the 11th day will find it relatively easy to relate to most other people. SODIUMs are often found to be inviting, likable, alluring, charitable and giving people. SODIUMs generally tend to have a higher need for social interaction with others. This can be expressed in varying ways such as having many friends; having the need to be in a relationship; or through socializing in groups/clubs or even by just being friendly with strangers. SODIUMs generally find it easy to mingle, make new friends, and seem to have a way of making even the coldest personalities open up to them. Though very outgoing and engaging, they are somewhat cautious and skeptical of others, yet it is rare for them to outwardly display or act on this without just cause. The close relations they form can be some of the deepest, most meaningful, and long-lasting in comparison to others. They are governed by a good balance of both emotion and intuition & thought and analysis; they can vacillate between utilizing their instincts and reasoning. They are typically very socially intuitive which provides them with the gift to quickly sum up strangers. This in turn, helps them to balance and overcome their suspicions of others. They can often lend a helpful ear and be a source to ease one's troubles and mind. They can be good at recognizing and hold the skills to effectively defuse high tension situations before they escalate. They can also be great planners, motivators, and helpful with moving things along when they fall behind, slow or come to a stop. SODIUMs can be interesting characters with a story to tell.

ADDITIONAL DEFINING QUALITIES, CHARACTERISTICS & INTERESTS

SODIUM is a very soft silver-white metal that is extremely malleable; has a very low density; and is an excellent conductor of heat and electricity. SODIUM is vital to animals and many plants. It floats on, and spontaneously ignites when exposed to water. SODIUM personalities can come across as very easy going, and perhaps they are perceived by some to be too easy going. However, they are not slouches or pushovers as they can handle their own. They can be very good listeners, observers, and general companions. They seem to have an inherent skill to make even the coldest personalities open up to them. SODIUMs also tend to be very flexible and accommodating. They prefer to keep things on the simple side and don't necessarily like or buy into the complexities of life. They prefer to avoid drama; are sentimental and can hold onto memories and feelings longer than they should. They are typically good at maintaining their cool but can be somewhat unpredictable when dealing with stressful situations. However, through their elevated social intuitiveness, they are good at recognizing the need for and or taking the time to maintain their health and peace of mind. This can be accomplished through various methods which leads them to have interests in such things as exercise, diet, meditation, yoga, socializing, or just taking time to decompress the best way they know how. SODIUMs are people oriented and also have a connection with nature - animals, the outdoors, and water. SODIUMs tend to like baking, cooking, cleaning, food, dining, caring for the environment, exercise, reading, writing, photography, clothes, architecture, maintaining peace of mind, medicine, the ocean, clothes, nightlife, music, libations, and their appearance/hygiene. They enjoy city life, and like to be "out and about" and "be seen" but they are not flashy or overly needy in this regard.

12 THE DAY OF MAGNESIUM

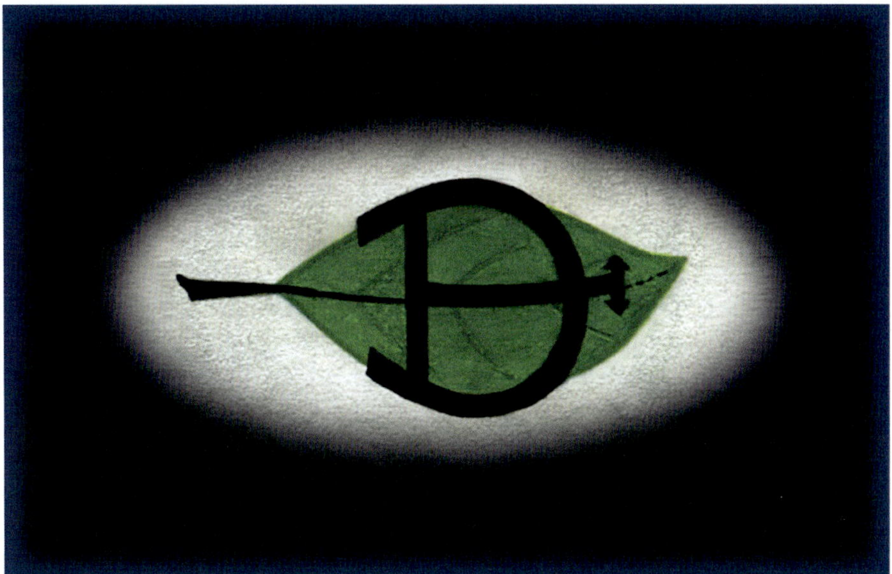

LIFE ELEMENT MAGNESIUM (Mg) -
12TH DAY BIRTH ELEMENT

CLASSIFICATION

MAGNESIUM is a solid metal at standard temperature and pressure but is highly flammable when spread out. Thus, it possesses both Earth and Fire Element qualities. It is classified as an Alkali Earth Metal. It is a good conductor of heat and electricity; has a low density, melting point and boiling point. It is reactive with water to form very strong and dangerous hydroxides.

COMPATIBILITY

MAGNESIUM personalities are very compatible with Life Element Oxygen (8) as that is the 2nd most abundant combination in the earth's crust. They also relate well with Hydrogen (1), Beryllium (4), Boron (5), Carbon (6), Sodium (11), Silicon (14), Phosphorus (15), Sulfur (16), Potassium (19), Calcium (20), and Scandium (21). They can also have very strong connections with Fluorine (9) and Chlorine (17). MAGNESIUM is usually an inspirational and overall positive presence for Aluminum (13), Vanadium (23), Chromium (24), Manganese (25), Iron (26), and Nickel (28): the combination of MAGNESIUM with any one of these typically results in the betterment for both of the Life Elements.

INCOMPATIBILITY

MAGNESIUMs may not necessarily have much in common with Helium (2), Lithium (3), Helium (10), Argon (18), Titanium (22), Cobalt (27), Copper (29), Zinc (30), and Gallium (31). Though MAGNESIUMs and Nitrogen (7) have an attraction and commonalities, they can also have many tensions between them and are often better off avoiding one another.

ELEMENT BACKGROUND, GENERAL DISPOSITION QUALITIES

MAGNESIUM ranks 9th in abundance in the Universe and 8th among all Life Elements on Earth. It is the 11th most common Life Element found in the human body and is essential for all life forms as we know them. MAGNESIUM is highly flammable when its surface area is widely spread out. MAGNESIUM is at the core of Photosynthesis. Without MAGNESIUM plants, algae, and cyanobacteria would not exist. In turn, animals, including humans, would not exist either. It is the lightest of all structural metals. People born on the 12th day tend to have many commonalities and will get along with many other Life Elements. They typically find it relatively easy to interact with most people. They are generally friendly and also find it easy to form friendships with most people. They can also be very altruistic people and will sacrifice their own resources, time, efforts, and own advancement for the betterment of others. MAGNESIUMs are governed more by their emotions and intuition over thought and analysis. MAGNESIUMs can be good at picking up and sensing the energies and vibes of others better than most. MAGNESIUMs enjoy the company of others and typically do not like to be alone. They tend to be relatively stable and calm and typically remain so when confronted with significant issues. Thus, they tend to handle stress moderately well. However, when spread too thin, over-stressed and or pushed too far they can become very emotionally fierce and intense. MAGNESIUMs can be their own worst enemy when they do not realize or appropriately manage conflict and stressors in their lives. Once MAGNESIUMs lose their cool it may be very difficult to calm them. The best resolution is to get out of their way and wait for them to cool down on their own.

ADDITIONAL DEFINING QUALITIES, CHARACTERISTICS & INTERESTS

MAGNESIUM is a silvery-white, very light, and strong metal. It has a very high thermal conductivity. Magnesium has the ability to tarnish which in turn protects it from rusting or decomposing further. MAGNESIUM is the basis of Photosynthesis. Without MAGNESIUM, photosynthesis would not exist and thus nearly all plants, trees, flowers, algae, and certain bacteria would not exist. It is also vital to all animals. In the human body alone it is known to be crucial for at least 300 biological processes. Though MAGNESIUM personalities can be very friendly and compatible, they may at the same time be skeptical of others they don't know and may tend to "keep their guard up". They also in turn tend to be cautious with whom they allow close to them. MAGNESIUMs can be flashy; strong willed, and sometimes stubborn, and at times even "fiery"; they can have strong personalities and be set in their ways. They are lovers of life. They love plants, flowers, trees, animals, Nature, the outdoors, the water, and the Sun. They are generous, especially with their time and hearts; they like technology; they can be good with their hands and appreciate crafts/arts. They like science, and have a particular fascination with health and medicine, space

& space exploration, and aeronautics/planes/flying. MAGNESIUMs can be somewhat concerned with their appearance versus others; they also tend to love food and cooking and may be vegans more often than any other Life Element as a result of their affinity to plants. They also may have interests in construction, architecture, cars, domestic tasks, photography, travel, the armed forces, and recreation.

13 THE DAY OF ALUMINUM

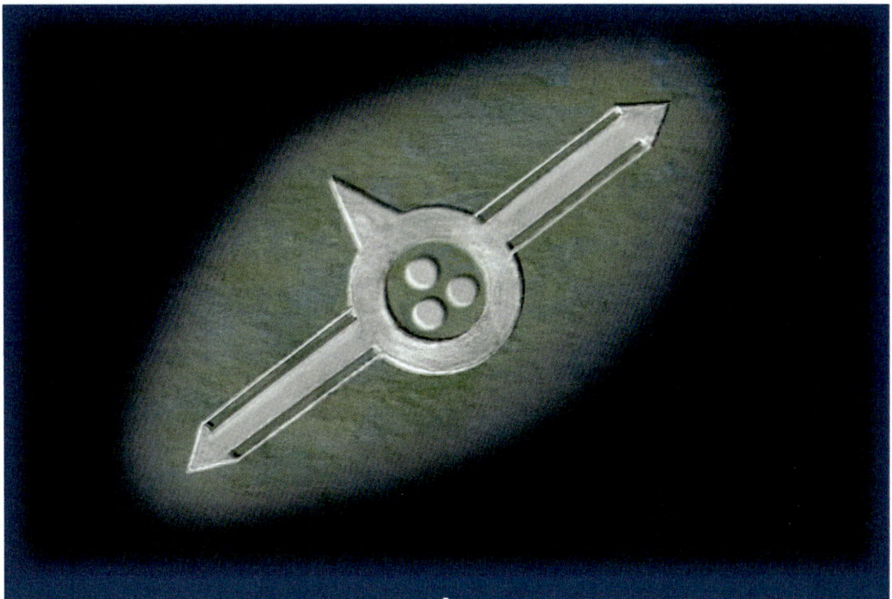

LIFE ELEMENT ALUMINUM (Al) - 13TH DAY BIRTH ELEMENT

CLASSIFICATION

ALUMINUM is a solid metal at standard temperature and pressure and thus possesses Earth Element qualities. It is classified as an Other Metal which means it utilizes its three outer electrons to form covalent bonds (it typically shares its electrons when forming bonds). It is very light, ductile and malleable.

COMPATIBILITY

ALUMINUM personalities are most compatible with Oxygen (8) and Silicon (14). They also relate well with Lithium (3), Beryllium (4), Carbon (6), Fluorine (9), Sodium (11), Magnesium (12), Phosphorous (15), Sulfur (16), Chlorine (17), Potassium (19), Calcium (20), Scandium (21), and Iron (26). ALUMINUMs can also have mutually beneficial relations with Titanium (22), Vanadium (23), Chromium (24), Manganese (25), Cobalt (27), Copper (29), Zinc (30), and Gallium (31).

INCOMPATIBILITY

ALUMINUM personalities are less likely to be compatible with Hydrogen (1), Helium (2), Boron (5), Nitrogen (7), Neon (10), Argon (18), and Nickel (28).

ELEMENT BACKGROUND, GENERAL DISPOSITION QUALITIES

ALUMINUM is at the high end of abundance in relation to other elements in the Universe and on Earth. It is the 14th most common element in the Universe and 3rd in the Earth's crust behind Oxygen (8) and Silicon (14). It is the 21st most abundant element found in humans. It is the most abundant metal found in the Earth's crust. It is widely dispersed and found in many different compounds. People born on the 13th day tend to easily find some common ground with most other Life Elements. Therefore, ALUMINUMs can find it pretty easy to relate to most other Life Elements. They are typically social, outgoing, versatile, and likable. ALUMINUMs can have very diverse social circles and interests, perhaps more so than any other Life Element. ALUMINUMs usually possess a wide skill or knowledge base and work best in groups, paired up or when teamed with others. ALUMINUMs are giving and like to share their knowledge, skills, and thoughts. Thus they make for wonderful leaders, managers, and teachers of all sorts, workmates, teammates, and partners. ALUMINUMs prefer the company of others and are relationship oriented people. The bonds they form are typically very strong and it is more difficult for them to end relationships with friends and romances in comparison to others. They are very good with handling regular day to day stress, but they are not necessarily the best at holding up under great stress or pressure over extended periods. However, in these very high-stress situations, they tend to keep themselves together well enough that most would not recognize the struggles they are experiencing. That is, ALUMINUMs can be a mess inside but not show it externally. Nonetheless, under prolonged high stress and pressure, they will eventually visibly show signs of physical distress. They strike a very good balance between being thoughtful and analytical versus being guided by emotion and intuition. They also have a nice mix of being cautious but open at the same time to others. ALUMINUMs are gifted and somewhat unique in the way that they can be highly perceptive of the "energy" that others emanate. ALUMINUMs "play well with others" and are typically fun people to be around.

ADDITIONAL DEFINING QUALITIES, CHARACTERISTICS & INTERESTS

ALUMINUM is a soft, durable, lightweight, ductile and malleable metal; it is silver-like with a slightly bluish tint. It is an excellent electrical and heat conductor. It is very reflective of both visible light and infrared radiation (non-visible energy from the Sun). ALUMINUM is very widely found throughout nature and in many man-made items. ALUMINUM has no known biological role for life on Earth as we know it. It is ingested in certain foods in small amounts. There is some evidence that it may play a role in certain brain and nervous system disorders such as Alzheimer's disease - Alzheimer patients have increased deposits of ALUMINUM in the brain. It is toxic to many plants when it breaks down in aqueous acidic soils. ALUMINUM has the ability to bond with many other elements. ALUMINUM is unique in that it reacts with both acids and bases - there are very few elements that do this. When exposed to Oxygen (8), ALUMINUM forms a protective layer around itself which prevents it from corrosion. ALUMINUMs most often prefer interaction with people. ALUMINUMs are capable of getting along with a wide variety of differing personalities. They are often open to learning and listening to all sides before passing judgment. ALUMINUMs generally tend to be rather introspective, do not let others get under their skin very easily, are reliable, and trustworthy. They like being "out and about" but they are more oriented toward being around people rather than their fondness of being outside. They also tend to enjoy dining, food, cooking, libations, music, shows, reading, travel, technology, crafts, sports, toys (cars, boats, planes, etc.) mechanics, construction, architecture, homes, doing home projects and home decor; they also enjoy games and brain teasers; and enjoy entertaining. They also tend to care about their appearance.

14 THE DAY OF SILICON

LIFE ELEMENT SILICON (Si) - 14TH DAY BIRTH ELEMENT

CLASSIFICATION

SILICON is a solid at standard temperature and pressure and thus possesses Earth Element qualities. It is classified as a Metalloid. As a Metalloid, SILICON has both metal and non-metal characteristics. It is one of only two Life Elements that are Metalloids, Boron (5) being the other. It typically shares its outer 4 electrons to form stable bonds.

COMPATIBILITY

SILICON personalities are most compatible with Oxygen (8). They also get along well with Lithium (3), Beryllium (4), Boron (5), Fluorine (9), Sodium (11), Magnesium (12), Phosphorus (15), Potassium (19), Calcium (20), Scandium (21), Titanium (22), Vanadium (23), Chromium (24), Manganese (25), and Zinc (30). They can also make wonderful mutually beneficial relations with Aluminum (13), Iron (26), and Copper (29).

INCOMPATIBILITY

SILICON personalities are less likely to be compatible with Hydrogen (1), Helium (2), Carbon (6), Nitrogen (7), Neon (10), Sulfur (16), Chlorine (17), Argon (18), Cobalt (27), Nickel (28), and Gallium (31).

ELEMENT BACKGROUND, GENERAL DISPOSITION QUALITIES

SILICON is the 8th most abundant Life Element in the Universe of known matter. It is the 2nd most common Life Element in the Earth's crust and the 12th in the human body. It is vital for life in both plants and animals. It has the ability to either give up or share its outer 4 electrons which in turn allows it to react with many other Life Elements and compounds to form a wide range of substances. SILICON and Oxygen (8) combine to make up the most abundant compound combination on Earth. The majority of stone, clay, and sand on Earth contains SILICON and Oxygen (8) compounds. SILICON exists in multiple forms. One form is shiny, grayish-black, and has needle-like crystals or is flat plate-like. The other common form is a brown powder. SILICON is likely the most widely manipulated and utilized Life Element by mankind even more so than Iron (26). It is very stable at standard temperature and pressure where it is considered to be inert (does not readily react with other elements). SILICONs are individuals who are generally caring, sensitive, complex, stable, and nostalgic. They possess many varying characteristics as a result of being a

Metalloid which can also make them rather unpredictable and very difficult to figure out. They can have a wide variety of interests and social circles, and can wear many hats; they are diverse and extremely adaptive. They are generally inquisitive (they like to know how things work) and enjoy pondering many of life's mysteries. They will put forth great effort to resolve problems or issues that are assigned to them or that have meaning in their lives. When they want, SILICONs can find it very easy to relate with most other Life Elements. They are passionate, love deeply, and can hold onto things probably longer than they should. They often appreciate multiple perspectives and strive to view all things from different viewpoints. SILICONs can often be very outgoing but can also just as easily be introverted. SILICONs like the company of others but it is usually most preferred on their own terms. They are innately relationship type people and are therefore at their best when paired with the right Life Element. Some SILICONs can remain exceptionally focused and calm during dire and critical situations; yet in normal day to day situations they can come across as aloof and uncertain of themselves or the situation they are in. Most SILICONs tend to handle stress better than most especially abrupt and intense stressors. They can also be good with day to day stress so long as they recognize their stressors and appropriately channel it. Though they tend to rely heavily on their emotions and intuitions, they are innately governed more and better served by thought and analysis over emotion and intuition. That is, like many, they can often initially react emotionally or judge based on intuition, but with discipline over time and when in tune with themselves they are best served by taming their initial reactions by focusing on thought and analysis versus their emotions and intuition. SILICONS are some of the most resolute, collected, and focused people when they need to be. They are "processors", meaning they survey and analyze nearly everything. They typically are not easily persuaded in general. However, that does not mean they are unwilling to be open to hearing and accepting varying opinions as these are usually sought out by them.

ADDITIONAL DEFINING QUALITIES, CHARACTERISTICS & INTERESTS

SILICON exists in its elemental form either as a grayish-bluish-black with needle-like crystals or flat plate-like texture; or as a brown powder. It is hard and is a semiconductor which means it is a moderate conductor of electricity. It has a relatively high melting point and thermal conductivity. It is relatively inactive under normal conditions: it will not react with air or most other elements, water, or steam; and most acids have very little effect on it. There is much to learn about its biological role and many ongoing studies are attempting to do just that. However, we do know that it is required for the synthesis of elastin and collagen; the aorta contains the highest content of elastin and SILICON. It has also been shown to improve cell wall strength and structural integrity in plants, improve drought and frost resistance, decrease lodging potential and boost the plant's natural pest and disease fighting systems. SILICON personalities can be perceived as straightforward, direct, yet having many layers. They can bring much passion to a relationship; they love deeply and make very strong connections. They can also have difficulty letting go of many things. They tend to be sentimental and as a result they may hold onto some things longer than most - these can be people, physical objects, mental thoughts or emotions/feelings. Many SILICONs enjoy sharing and engaging with others in many ways; they appreciate reciprocation of giving and taking whether it be material, physical or emotional. They can bring structure where it is lacking. They tend to be rather active and have a wide range of interests such as cooking, baking, the arts, crafts, music, reading, electronics (computers, phones, audio, etc.), cars, plants, flowers, nature, outdoors, animals, health/fitness/medicine, playing games/toys (brain teasers, electronic, board, etc.), architecture, anthropology, construction, large buildings and bridges, travel, boats, time/aging, ancient and modern cultures, and history. SILICONs can also have an affection for cleanliness and caring about their appearance.

15 THE DAY OF PHOSPHORUS

LIFE ELEMENT PHOSPHORUS (P) - 15TH DAY BIRTH ELEMENT

CLASSIFICATION

PHOSPHORUS is a solid at standard temperature and pressure. However, it also spontaneously ignites in air and thus possesses both Earth and Fire Element qualities. It is one of the six “basis of life” elements - CHONPS (Carbon, Hydrogen, Oxygen, Nitrogen, PHOSPHORUS, and Sulfur). It is classified as a Non-metal. PHOSPHORUS, like Nitrogen, has 5 electrons in its outer shell and it can form three bonds with three different elements or less at the same time - this is somewhat unique.

COMPATIBILITY

PHOSPHORUS personalities are most compatible with Oxygen (8), and Calcium (20). They also relate well with fellow PHOSPHORUSs, Hydrogen (1), Lithium (3), Beryllium (4), Nitrogen (7), Fluorine (9), Sodium (11), Magnesium (12), Aluminum (13), Silicon (14), Sulfur (16), Chlorine (17), Scandium (21), Chromium (24), Manganese (25), Iron (26), Nickel (28), Copper (29), and Zinc (30).

INCOMPATIBILITY

PHOSPHORUS personalities are less likely to be compatible with Helium (2), Boron (5), Neon (10), Argon (18), Potassium (19), Titanium (22), Vanadium (23), Cobalt (27), and Gallium (31). Although they are attracted to one another Carbon (6) and PHOSPHORUS are not necessarily the best fit in the long run. Also, highly attracted are Fluorine (9) and PHOSPHORUS but they can have destructive associations and thus may be better avoiding one another.

ELEMENT BACKGROUND, GENERAL DISPOSITION QUALITIES

PHOSPHORUS is at the high end of abundance in the Universe and in the middle on Earth with respect to other elements. It is the 18th most common element in the Universe, 13th on the Earth's surface and 6th in the human body. PHOSPHORUS is widely dispersed on Earth and found in many different compounds. It is unique in that it is essential for all life organisms, among many other reasons, it is exceptionally unique given it is the only "fuel" source for ***all*** living cells in ***all*** living things as we know them. Compared to most, PHOSPHORUSs can find it pretty easy to relate to many other Life Elements. However, there are some personalities that they will struggle to find common ground or compatibility with. They are generally social, outgoing, and inviting. PHOSPHORUSs can exhibit diverse interests in relation to other Life Elements. Though PHOSPHORUSs do not always strike commonalities with everyone, they can still have a pretty diversified group of friends. PHOSPHORUSs can be somewhat unpredictable because they can possess contradictory characteristics. They can relate to and understand both extremes of many issues and topics; they can also be very good at relating to both sides of an argument or disagreement. They are generally content with themselves and who they are. They typically respond well to change compared to others. Many PHOSPHORUSs are energetic, and some can be over intense in their day to day demeanor. Others will be very calm, yet passionate, but they too under certain circumstances will exhibit over-impassioned behavior. PHOSPHORUSs are generally very giving and love to interact and share with others. They can be some of the most giving and sharing people. They can be known to have glowing personalities. They are generally people persons that have an extreme appreciation for and connection with Nature, its inhabitants and the natural world in general. PHOSPHORUSs prefer the company of others and are relationship-oriented people. However, they typically are not needy in this regard and though it may not be their preference, they can be content by themselves from a romantic perspective. Additionally, they may not feel as lonely as most when they are alone. This may be a result of them being innately content coupled with the fulfillment they achieve through their diverse social network and interactions with friends. They are not very good at handling stress. PHOSPHORUS personalities need to take extra caution and heed close attention to how they process and manage stress as it can make them extremely volatile which will affect them mentally and physically, and like any of us, it can eventually be a threat to their peace of mind over time. They are typically governed by their emotions and intuition over thought and analysis. They will rely more on their instincts when assessing others and approaching potential relations of all kinds. PHOSPHORUSs are capable of getting along with a wide variety of differing personalities despite their inherent differences with some Life Elements. PHOSPHORUSs can be the "life of the party".

ADDITIONAL DEFINING QUALITIES, CHARACTERISTICS & INTERESTS

PHOSPHORUS in its elemental state exists in multiple forms. The two most common types are White PHOSPHORUS which is white, transparent, waxy, toxic and unstable; and Red PHOSPHORUS which is red amorphous (shapeless, powder-like) is non-toxic and more stable. PHOSPHORUS is essential for all life on earth as we know it. It is incredible in the way that it is the Life Element that supplies ***all*** energy to ***all cells*** in ***all*** living organisms in Nature - it is the basis of all fuel sources in Nature. Without PHOSPHORUS all living things would die in seconds. It also plays a major role in the structural framework of all cell membranes and of DNA and RNA molecules. Additionally, it is essential for our bones, teeth and is used to build complex molecules, contract muscles, generate electricity in nerves, and is the source of light we see in fireflies. PHOSPHORUSs are often active, energetic and love socializing with people. They are well suited to be the driving force behind whatever needs to be done. Motivation, comradery, and appeasement are characteristic of PHOSPHORUS personalities. Despite their appreciation and fondness for connecting with people they also enjoy their alone time too. They have a deep, call it, spiritual connection with Nature including animals, plants and all of life. They often enjoy most anything outdoors, recreation, gardening, electronics, visual stimuli, cooking, cleaning, staying active, hygiene, flowers, pets, entertaining/hosting, travel, medicine, science, and the water.

16 THE DAY OF SULFUR

LIFE ELEMENT SULFUR (S) - 16TH DAY BIRTH ELEMENT

CLASSIFICATION

SULFUR is a solid under normal conditions and thus possesses Earth Element qualities. It is one of the six "basis of life" elements - CHONPS (Carbon, Hydrogen, Oxygen, Nitrogen, Phosphorous, and Sulfur). It is classified as a non-metal which means it is not a good conductor of heat or electricity.

COMPATIBILITY

SULFUR personalities are most compatible with Life Elements Hydrogen (1), Carbon (6), Nitrogen (7), Oxygen (8), Fluorine (9), Sodium (11), Magnesium (12), Aluminum (13), Phosphorus (15), Chlorine (17), Argon (18), Potassium (19), Calcium (20), Vanadium (23), Chromium (24), Manganese (25), Iron (26), Cobalt (27), Nickel (28), Copper (29), Zinc (30), and Gallium (31). SULFUR also has quite an attraction for other SULFURs.

INCOMPATIBILITY

There are very few Life Elements that SULFUR does not react with. Nonetheless, SULFURs are least compatible with Helium (2), Lithium (3), Beryllium (4), Boron (5), Neon (10), Silicon (14), Scandium (21), and Titanium (22).

ELEMENT BACKGROUND, GENERAL DISPOSITION QUALITIES

SULFUR is at the high end of abundance in the Universe and in the middle on Earth with respect to other elements. It is the 10th most common element in the Universe, 15th on the Earth's surface and 7th in the human body. SULFUR is one of the most reactive Life Elements yet it is found in great abundance in its elemental form on Earth. This is an unusual characteristic because elements that are highly reactive are typically not found in their elemental state. It forms bonds with other Life Elements in multiple ways. It can share its outer 6 electrons with other elements, or it can give up or receive electrons when bonding. It is widely spread throughout the planet and forms many different compounds. It forms stable compounds with nearly all other Life Elements; some of these compounds form at standard temperature and pressure while others at non-standard conditions. People born on the 16th day can find it easier than most to get along with many different people. They can be very happy and content by themselves just as much as they can be with others. This is not to minimize their love for being with others as they are certainly people persons. When it suits them, they are capable of easily making friendships and can form quick long lasting bonds with many different personality types. SULFURs also are capable of taming some of the strongest personalities. They are innately giving; open; like to share, and enjoy receiving back from their sharing efforts. They are also often genuine, deep, laid-back, well natured, and make for wonderful friends. They are very emotionally driven and thus may tend to rely more on their instincts and intuition versus their thoughts and analysis despite being "thinkers". That is, they are very analytical and deep thinkers regarding many topics but their emotional side will usually win over in their judgment and reactions regarding people and situations. They can be one of the most resourceful personalities; that is, they can be very creative to solve any problem that confronts them. This is especially true when they are seeking any type of interaction with, input, or cooperation of others. SULFURs, somewhat tend to not let things get to them too much. However, when rubbed the wrong way by certain individuals some SULFURs can exhibit fierce responses despite their general tendencies to avoid conflict. They process and handle stress pretty well on a day to day basis. However, when they are confronted with an on-going elevated stressor(s) they do not do so well in comparison to others. They must take the necessary actions to deal with these elevated and prolonged stressors before they are disturbed by them to a point that causes serious disruptions or problems in their lives and those around them. They are more often "lover not fighter" types.

ADDITIONAL DEFINING QUALITIES, CHARACTERISTICS & INTERESTS

SULFUR is a non-toxic, tasteless, odorless, brittle solid that is pale yellow in color, and a poor conductor of electricity. SULFUR is a necessity for all living cells and thus terrestrial life including plants and animals. It has many uses in biochemical processes such as fuels, vitamins, enzymes, antioxidants, and proteins. SULFUR is unique in that it forms very stable octatomic (8 different SULFUR atoms bond together) compounds with itself. SULFUR personalities relate very well with others of the same Life Element (SULFUR). They enjoy being with others and tend to have a high need for companionship. This desire to be with others can be satisfied through many avenues such as friends, romance, clubs, social media and groups. They are very accommodating to others, more so than most others. In this regard, they can be very creative and will go far beyond what most will do to ensure that others feel welcomed, needed, or satisfied. They will also go to extremes to ensure those who they seek the attention of notice them. They tend to normally be calm, easy going, and effortlessly content. However, though they prefer to avoid confrontation, they can unexpectedly be very decisive, and take charge of a situation or circumstance when required. SULFUR personalities tend to have an affection for all of Nature and all of its living inhabitants. They also tend to enjoy the outdoors, plants, animals/ pets, fruits & vegetables, wilderness, flowers, reading, clothing, geography, social clubs of many sorts, entertaining, social events (concerts, shows, etc.), wine, jewelry, aesthetics (art, natural beauty, etc.), medicine, alternative/natural healing, health and fitness, and are foodies. They can also often be concerned with the cleanliness of themselves and their surroundings.

17 THE DAY OF CHLORINE

LIFE ELEMENT CHLORINE (Cl) - 17TH DAY BIRTH ELEMENT

CLASSIFICATION

CHLORINE is a gas under standard conditions and thus possesses Air Element qualities. CHLORINE is one of two Halogen Life Elements, Fluorine (9) being the other. The Halogens are extremely reactive. In fact, they are the most reactive of all the elements. CHLORINE will typically take an electron when bonding but it can also give up anywhere from 1 to 7 of its outer shell electrons too.

COMPATIBILITY

CHLORINE personalities are most compatible with Sodium (11). They can also get along well with Life Elements Hydrogen (1), Lithium (3), Beryllium (4), Boron (5), Carbon (6), Nitrogen (7), Magnesium (12), Aluminum (13), Phosphorous (15), Sulfur (16), Potassium (19), Calcium (20), Vanadium (23), Iron (26), Copper (29), and Zinc (30). They also have an attraction to Lithium (3) but may not be the best long-term fit.

INCOMPATIBILITY

CHLORINE is the 2nd most reactive Life Element - Fluorine (9) is the most reactive. It reacts with nearly all other Life Elements under certain conditions. However, CHLORINEs are least compatible with Helium (2), Oxygen (8), Neon (10), Silicon (14), Argon (18), Scandium (21), Titanium (22), Chromium (24), Manganese (25), Cobalt (27), Nickel (28), and Gallium (31). Though they can under certain circumstance get along well, CHLORINEs and Hydrogens (1) can also have rocky relations and thus may not necessarily be best suited for one another despite their attraction. CHLORINEs may also be better off avoiding Fluorines (9) too despite having many similar characteristics. CHLORINEs are also very attracted to Lithiums (3), but they may not always be best long-term as these relations can sometimes be volatile.

ELEMENT BACKGROUND, GENERAL DISPOSITION QUALITIES

CHLORINE ranks 23rd in abundance in the Universe amongst all elements. It is very common on Earth and is the 3rd most common element in the oceans. It is the 9th most common Life Element in the human body. CHLORINE is extremely reactive and easily combines with nearly all the elements at standard temperature and pressure. Other than in very small amounts in volcanic gasses, it is rarely found on Earth in its elemental state. People born on the 17th day typically find it easy to engage and interact with others; they have an inherent ability to connect with others. They can often be alluring, somewhat seductive, and irresistible to some. They are perhaps the most unrestrained Life Element by nature and thus they can easily interact with others. This allows them more opportunities to form many friendships and romantic bonds. However, CHLORINEs also tend to have contentious interactions with many other Life Elements both in their day to day relations and in romance. They are less likely to have a "filter" which can lead to them leaving some that they meet with a not so favorable impression. CHLORINEs tend to have very strong, deep, and passionate romantic relationships despite the tumultuous interactions they may have with some. They similarly have deep and purposeful friendships. They are loyal and can be to a fault. They are very emotionally and intuitively driven by nature versus being analytical and thoughtful. CHLORINEs also tend to be very accepting, understanding and open to others. Though they will have little issue speaking their mind, CHLORINE personalities can have challenges with identifying and expressing their own innermost thoughts and feelings. However, they can be very good listeners. They are typically not the best when it comes to dealing with stress. They also tend to be a source of and or associated with drama more than most.

ADDITIONAL DEFINING QUALITIES, CHARACTERISTICS & INTERESTS

CHLORINE is a highly toxic pale yellow-green diatomic gas in its natural state. Though it combines with itself, the bond between the two CHLORINE atoms is very weak thus making it unstable and highly reactive. Though it is toxic, it is also a necessity for all life forms in low-level concentrations. CHLORINE is utilized by the body in numerous ways. It is used to form hypochlorite as part of the immune response by the body to kill bacteria; it is also found in the stomachs of animals in the form of hydrochloric acid (HCL) which is used to aid in the breakdown and digestion of food. CHLORINE personalities can have some commonalities with other CHLORINEs. They require and enjoy some alone time. However, the alone time CHLORINEs require is short lived. CHLORINE personalities have a very high need for interaction and companionship and will become extremely restless when they are alone for too long. CHLORINEs can be friendly and outgoing, but they also can have much drama and turbulent relations. They tend to have to work harder than most in order to maintain an amiable disposition and relations with others. Despite the difficulty they may have with others, they are people persons, yet they sometimes seem to thrive on drama. Some Life Elements, such as Sodium (11) and Potassium (19), can be very good counterbalances for subduing CHLORINEs dramatic tendencies. CHLORINEs love water and also tend to appreciate cleanliness, cars, home décor/interior decorating, plants, gardening, cooking/dining/food, reading, art, music, photography, construction, cars, the ocean and all water in general, the environment, animals/pets, games, intimacy, and the outdoors. CHLORINE personalities also have a very high need for companionship, and therefore tend to belong to varying types of groups or clubs, engage heavily in social media, are in a relationship or seek one more than most.

18 THE DAY OF ARGON

LIFE ELEMENT ARGON (Ar) - 18TH DAY BIRTH ELEMENT

CLASSIFICATION

ARGON is a gas under standard temperature and pressure and thus possesses Air Element qualities. It falls into the category of Noble Gases. As a Noble Gas, its outer ring is filled to the limit with electrons. Therefore, it is very stable and does not readily form compounds with itself or any other element.

COMPATIBILITY

ARGON personalities are most compatible with and tend to get along best with Hydrogen (1), Helium (2), Nitrogen (7), Oxygen (8), Fluorine (9), Neon (10), sulfur (16), Potassium (19), Titanium (22), and Gallium (31). They also get along with other fellow ARGONs.

INCOMPATIBILITY

ARGON, like Helium (2) and Neon (10) do not react directly with many if any other Life Elements. Lithium (3), Beryllium (4), Boron (5), Carbon (6), Sodium (11), Magnesium (12), Aluminum (13), Silicon (14), Phosphorus (15), Chlorine (17), Calcium (20), Scandium (21), Vanadium (23), Chromium (24), Manganese (25), Iron (26), Cobalt (27), Nickel (28), Copper (29), and Zinc (30) are likely to have less in common with ARGON compared to other Life Elements.

ELEMENT BACKGROUND, GENERAL DISPOSITION QUALITIES

ARGON is the 11th most common element in the Universe. It is the 3rd most common element in the atmosphere and is found in very low concentration in the Earth's crust. It is not found in the human body. ARGON is continually created on Earth by the decay of Potassium (19). ARGON is very stable and it does not react with any other element under normal conditions. ARGON is not found in living beings on Earth, or anywhere else on Earth that we are aware of. More than most, people born on the 18th day can have their difficulties relating with many of the other Life Elements. They can find it tough to form close bonds with many other people. They will often form friendships, but compared to most others, many of them are often shallow or not necessarily deep and long-lasting. However, ARGONs may be partly to blame for this as they, in comparison to most others, can be more content with being with themselves at times and a select few otherwise. They can be outwardly friendly but they will often keep much to themselves. As mentioned, ARGONs are more content on their own compared to other Life Elements. However, they do enjoy being social and fit in very well with the crowd despite their hesitation or difficulties forming close friendships with many Life Elements. ARGONs can be very good liaisons and mediators, and can quickly assess hostile situations and react accordingly. They are typically driven far more by thought and analysis versus their emotions and intuition. They tend to see through and/or avoid drama. They are excellent with processing and handling stress so long as they maintain their rituals that keep them in check. In fact, they are near, if not atop all Life Elements in this category. They are often level-headed, focused, successful, intelligent, goal oriented, straightforward, and "no nonsense" people. They can be very good at communicating; excluding their emotions and feelings which they will often struggle with expressing. They tend to be non-judgmental, don't necessarily take sides, are not easily swayed, and can sometimes come across as having dry personalities.

ADDITIONAL DEFINING QUALITIES, CHARACTERISTICS & INTERESTS

ARGON is a colorless, odorless, tasteless, non-toxic, and non-flammable gas. It is a relatively dense gas and has an extremely low melting point. It has no known biological need. ARGON people can be very open and transparent. As mentioned, ARGONs are generally very straight-forward people. They tend to be laid back; they don't excite easily; they are very protective and reserved. And though they are very thoughtful and calculated and can come across as dry; they can be very sentimental, cherish and hold onto to experiences and relationships. ARGON personalities tend to require alone time, and are also private people perhaps more than most. They tend to be very inward with their emotions and keep things to themselves. ARGONs can often settle with who they have relations with in times of weakness more than most. Unfortunately, this quality can give rise to the odds that they may interact with others that they perhaps were better off avoiding. ARGONs who are in tune with and maintain their inner instincts in difficult times will resist the temptation of settling or mingling with those that they should not. Like everyone, they need companionship, but ARGONs can also live happily on their own and be socially satisfied through just a few friendships, family members, or the social activities they engage. ARGONs tend to appreciate aesthetics, the arts, and can be concerned with their appearance. They also enjoy being outdoors; and have interests in science, medicine, electronics, technology, photography, movies, shows, concerts, the ocean, dining, cooking, wine/libations, and nightlife.

19 THE DAY OF POTASSIUM

LIFE ELEMENT POTASSIUM (K) - 19TH DAY BIRTH ELEMENT

CLASSIFICATION

POTASSIUM is an Alkali Metal. It is a solid at standard temperature and pressure and thus possesses Earth Element qualities. POTASSIUM also possesses Fire Element characteristics as well because of its tendency to spontaneously ignite when exposed to water. It typically gives up its outer electron to form ionic bonds. It is malleable, ductile, softer than most metals, and is an excellent conductor of heat and electricity.

COMPATIBILITY

POTASSIUM personalities are most compatible with Sodium (11), Magnesium (12), Sulfur (16), and Chlorine (17). They also get along well with Hydrogen (1), Boron (5), Carbon (6), Nitrogen (7), Oxygen (8), Fluorine (9), Aluminum (13), Silicon (14), Sulfur (16), Argon (18), Calcium (20), Chromium (24) and Manganese (25). They can also relate well with fellow Alkali Metal Lithium (3).

INCOMPATIBILITY

Compared with other Life Elements POTASSIUM personalities are less likely to be compatible with Helium (2), Beryllium (4), Neon (10), Phosphorus (15), Scandium (21), Titanium (22), Vanadium (23), Iron (26), Cobalt (27), Nickel (28), Copper (29), Zinc (30), and Gallium (31). Though POTASSIUM may have an attraction with Chromium (24), they may not always have the best relations.

ELEMENT BACKGROUND, GENERAL DISPOSITION QUALITIES

POTASSIUM is not one of the most common elements, but it is by no means rare in the Universe as it ranks 21st in abundance of all elements. It ranks as the 8th most abundant Life Element on the surface of the earth, and in the human body it is the 7th most common. POTASSIUM is one of the most reactive metals and is not found in Nature in its elemental state as a result of its high reactivity. People born on the 19th day will generally find it somewhat easier than most to interact with most other people. They relate well with many other Life Elements. However, there are just as many other personalities that POTASSIUMs will not get along with in the long run. They do not necessarily have bad relations with those that they eventually part with but they more so just end up not maintaining a meaningful connection. POTASSIUMs in general will often present themselves as inviting and likable people. In comparison, POTASSIUMs tend to have a higher need for social interaction with others. POTASSIUMs commonly find it somewhat easy to mingle and meet new people. Though very approachable and inviting, they are somewhat guarded and cautious when getting to know others on a deeper level. They are governed more by emotion and intuition over thought and analysis. They are typically very loyal and will give their very all and last effort for those who mean the most to them. They are also typically very giving and charitable individuals especially to those closest to them. They very much enjoy being with people and thrive on relationships; they oftentimes are found to be relationship type people as they more regularly require a relationship compared to others.

ADDITIONAL DEFINING QUALITIES, CHARACTERISTICS & INTERESTS

POTASSIUM is a very soft and light, silvery-gray metal that is extremely malleable; has a very low density; and is an excellent conductor of heat and electricity. In its elemental state, it tarnishes quickly when exposed to air as it reacts with Oxygen (8) to form a protective layer. It is the 2nd lightest metal behind Lithium (3). POTASSIUM is vital to all life as we know it. POTASSIUM personalities are typically easy going, laid back, light hearted and go with the flow type people. They are usually pretty outgoing and open people but can be somewhat guarded and skeptical of others, especially when it comes to allowing others into their lives in a meaningful way. They try not to allow the complexities of life to convolute them; they prefer simplicity and to keep things easy. POTASSIUMs also tend to be giving, flexible, and accommodating. As mentioned, they prefer to keep things on the simple side and don't necessarily like or buy into the complexities of life. With that said, they prefer to avoid drama. They are not very good at handling stress so their innate tendencies to keep things simple and avoid drama is quite helpful with maintaining their inner peace. POTASSIUMs are people oriented but also have a connection with nature - animals, the outdoors, and the wilderness. POTASSIUMs tend to like cooking/baking/dining, food, fruits/veggies, plants/flowers, recreation, the environment, TV/movies, music, pets, exercise, maintaining peace of mind, libations, medicine, the ocean/lakes/etc., and their appearance. They are also often health conscious, clean, and enjoy keeping their minds busy.

20 THE DAY OF CALCIUM

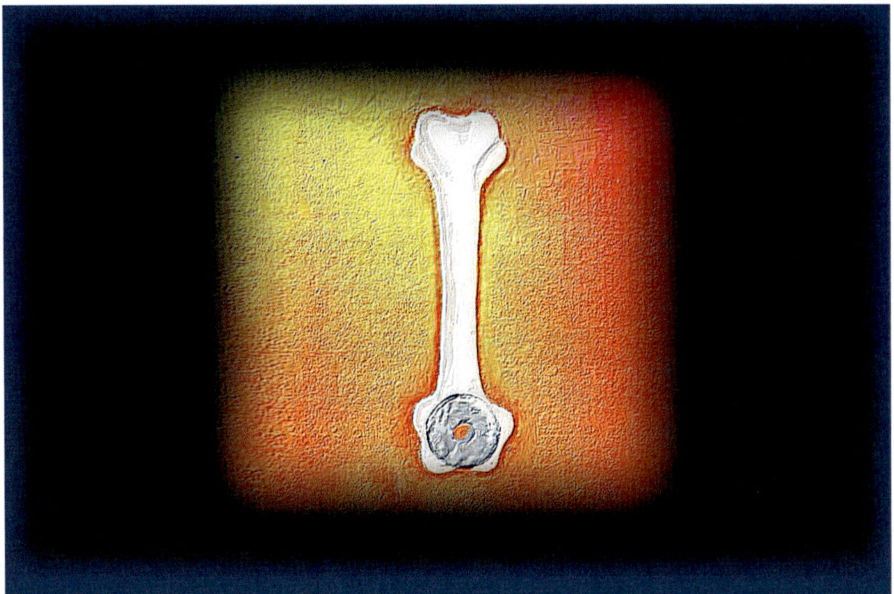

LIFE ELEMENT CALCIUM (Ca) - 20TH DAY BIRTH ELEMENT

CLASSIFICATION

CALCIUM is an Alkali Earth Metal. It is a solid at standard temperature and pressure and thus possesses Earth Element qualities. It is very reactive and typically gives up its 2 outer electrons when it forms bonds. It is an excellent conductor of heat and electricity. It has a very low density - it is the lightest Alkali Earth Metal; and has a high boiling point with a moderate melting point.

COMPATIBILITY

CALCIUM personalities are most compatible with Carbon (6), Oxygen (8), Sulfur (16), and Fluorine (9). They also tend to get along well with Boron (5), Nitrogen (7), Sodium (11), Silicon (14), Phosphorus (15), Chlorine (17), Potassium (19), and Titanium (22). They also relate to a lesser degree with Vanadium (23), Manganese (25), Iron (26), Cobalt (27), and Zinc (30). They can also relate with fellow Alkali Earth Metals Beryllium (4) and Magnesium (12). Additionally, they also can form very mutually positive relations with Aluminum (13) and Copper (29).

INCOMPATIBILITY

CALCIUM personalities are less likely to be compatible with Hydrogen (1), Helium (2), Lithium (3), Neon (10), Argon (18), Scandium (21), chromium (24), Nickel (28), and Gallium (31).

ELEMENT BACKGROUND, GENERAL DISPOSITION QUALITIES

CALCIUM is the 6th most abundant element in the Universe and 5th on the Earth's surface. It is also the 5th most common Life Element in the human body. CALCIUM is created from explosions of massive stars at the end of their lives. There are many known compounds on Earth that contain CALCIUM, many of which are very common. CALCIUM is an essential component of bones and teeth. It is also essential for all living organisms as it is used as a signal for many cellular processes. People born on the 20th day will typically find it easier than most to relate with other people in some way. They get along with many other Life Elements, more so than most. CALCIUMs are often very supportive and giving people but can sometimes be perceived as being distant and closed. However, CALCIUMs generally very much enjoy the company of others; they thrive in groups and interacting with others. They can be strong, firm, and dependable in their work ethic and their personalities in general. They can be the glue that keeps things together whether it be social groups, family, work projects, or personal one on one relations. They will champion causes for the greater good; altruism is often a characteristic of CALCIUMs. They also can have the ability to get those who are typically standoffish to open up. They are governed by emotion and intuition over analysis and thought. They can be high maintenance and on occasion be both the source of drama and engage in it. However, drama is typically not a regular day to day characteristic for CALCIUMs. They are also typically slightly below average when it comes to dealing with stress.

ADDITIONAL DEFINING QUALITIES, CHARACTERISTICS & INTERESTS

CALCIUM personalities require human interaction; like most, they do not like to be alone. CALCIUM forms a protective layer around itself with both Nitrogen (7) and Oxygen (8) when exposed to air. Thus, CALCIUMs can tend to be somewhat more picky, guarded and cautious with whom they allow close to them and who they romantically bond with. Though with Nitrogen (7) and Oxygen (8) they might not be able to help themselves.

CALCIUM is a soft silvery-gray metal that is an excellent conductor of heat and electricity. It possesses the 2nd highest thermal conductivity of the Life Elements behind Copper (29). It is the lightest of the Alkali Earth Metals. Pound for pound it is the best electrical conductor even more so than Aluminum (13) and Copper (29); however, its practical use is hindered by its reactivity with air. It has very few uses in its elemental form but its compounds are various and widely used. CALCIUM is essential for all life as it is needed in many cellular processes, and is a main component of leaves, bones, teeth, and shells. It reacts with water especially in powder form; it also reacts vigorously with some acids. CALCIUM personalities can be somewhat flexible; can be very good at planning; are efficient and detailed; and can also be very reliable. They are often very good at conceptualizing and implementing ideas. CALCIUMs can also be perceptive in such a manner that they can pick up on the "energy" of others; call it a sixth sense. They also tend to be more outward with expressing their emotions and do not keep things to themselves. They often are counted on to be the "go to" person. They also tend to be determined and resolute when there is a goal to be achieved. CALCIUMs also stand up well under great pressure especially when they have a supportive cast. They are dependent on others in their lives to funtion normally. They will be at their best when they have a romantic partner, supportive friends, and an active social life. They are both people oriented and have a love for Nature. CALCIUMs tend to have an affinity for the arts, theatre, travel, cars, sweets, plants/flowers, baking/cooking, socializing, health and medicine, the water, architecture/construction, and animals. They can also take extra care in their appearance and like to be noticed.

21 THE DAY OF SCANDIUM

LIFE ELEMENT SCANDIUM (Sc) - 21ST DAY BIRTH ELEMENT

CLASSIFICATION

SCANDIUM is a solid at standard temperature and pressure and thus possesses Earth Element qualities. It is classified as a Transition Metal. It is ductile, malleable; and conducts heat and electricity. It easily forms stable compounds with neutral molecules. It typically gives up 3 electrons when it bonds. Elements in the Transition Metal group are unique in the way that they can utilize more than one shell of electrons to combine with other elements. This allows them to form diverse and complex bonds.

COMPATIBILITY

SCANDIUM personalities are most compatible with Life Elements Beryllium (4), Silicon (14), Titanium (22), and Iron (26). They also get along with Oxygen (8), Magnesium (12), Phosphorus (15), and Manganese (25). Additionally, SCANDIUMs can have very mutually positive relations with Aluminum (13).

INCOMPATIBILITY

Hydrogen (1), Helium (2), Lithium (3), Boron (5), Carbon (6), Nitrogen (7), Fluorine (9), Neon (10), Sodium (11), Sulfur (16), Chlorine (17), Argon (18), Potassium (19), Calcium (20), Vanadium (23), Chromium (24), Cobalt (27), Nickel (28), Copper (29), Zinc (30) and Gallium (31) may not the best fits for SCANDIUM. Although SCANDIUMs are highly attracted to Oxygen (8), they can often have negative relations.

ELEMENT BACKGROUND, GENERAL DISPOSITION QUALITIES

SCANDIUM is not very common in the Universe in relation to other Life Elements. It is also not found in the human body. Though it is considered a rare earth element, it is commonly found in many different minerals in low concentrations that are widely dispersed throughout the planet. It is much more abundant in the Sun and other stars. SCANDIUMs tend to be generally calm and stable in comparison to others. They can also be quiet, reserved, and have relatively low drama with most people. However, be warned that they are no pushovers; when pushed into a corner or over pressured they can be fiery. They tend to steer away from drama and the crowd; and likely will not be found getting caught up in the latest fads. However, they can have an appreciation for the finer material possessions that our Earth has to offer. SCANDIUMs will typically find it difficult to relate with many other personality types compared to others. Though they do not connect as easily with others compared to most, they still are often likeable and can leave a lasting impression despite their reserved nature. This is especially true for certain other Life Elements that will find them hard to forget. They are also somewhat sensitive; and they can be very giving and accommodating to others. They are typically pretty good at processing and dealing with ordinary day to day stress. However, they can have some issues over time with long drawn out stressors that they are unable to avoid. They can also have issues with processing too much at one given time as they can be susceptible to being "scattered" and overwhelmed. SCANDIUMs have a very good balance when it comes to their assessment of others and their reactions to situations. They utilize a very good mix of their emotions and intuition along with thought and analysis. However, they are slightly more governed by their emotions and intuition over their thought and analysis methods. SCANDIUMs are typically pretty level-headed and generally work well with others as a result of their low drama, propensity to avoid conflict, and their no nonsense approach.

ADDITIONAL DEFINING QUALITIES, CHARACTERISTICS & INTERESTS

SCANDIUM is a soft metal with a silvery white appearance in its elemental state. It is malleable, ductile; it is fairly stable in air and reacts very slowly with some acids; it also has a very high melting point and is about as light as Aluminum (13). It is mainly used to make alloys with Aluminum (13) but it has a wide variety of applications. SCANDIUM interacts relatively calmly with most of the Life Elements that it combines with. They also have a tendency to form complex compounds. SCANDIUMs are relatively stable persons; they can work well with others; they don't mind taking direction from others and can be willing to perform tasks handed down to them - they can be content being the "worker bee". They typically are satisfied with who they are, their position, and place. They value the people and relationships in their lives more than anything such as family, friendships, lovers, neighbors, and their colleagues. SCANDIUMs can be inward with their emotions; they are not necessarily very expressive with their inner thoughts and feelings but are not reluctant to open up with the people they are comfortable with. Some SCANDIUMs may seem simple but they are often much deeper than the way they come across; they are typically introspective, complex, and can find themselves to be very analytical; and pondering & questioning many things in our world and beyond. As mentioned, they are also pretty calm people but they are not afraid to stand up for themselves or others that they care for. Despite their amiable disposition, they can also at times be high maintenance and difficult, though this is typically a minor part of their personality that will usually only surface from time to time or in certain individuals. SCANDIUMs commonly enjoy the outdoors, the Sun, and many types of outdoor recreation (camping, trails, shooting ranges, hunting/fishing, hiking, etc.). They also enjoy watching/playing many types of sports, plants/flowers/wilderness, travel, space/universe, science, technology, the armed forces, theatre, travel, thought provoking games/puzzles/conversation, and TV/movies.

22 THE DAY OF TITANIUM

LIFE ELEMENT TITANIUM (Ti) - 22ND DAY BIRTH ELEMENT

CLASSIFICATION

TITANIUM is a solid metal at standard temperature and pressure and thus possesses Earth Element qualities. It is classified as a Transition Metal. It is very ductile and is a poor conductor of heat and electricity. It typically shares electrons when it bonds. Elements in the Transition Metal group are unique in the way that they can utilize more than one shell of electrons to combine with other elements. This allows them to form diverse and complex bonds.

COMPATIBILITY

TITANIUM personalities are most compatible with Life Elements Silicon (14), and Calcium (20). They also get along well with Lithium (3), Boron (5), Sodium (11), and Scandium (21). TITANIUMs also make mutually beneficial relations with Aluminum (13), Argon (18), Vanadium (23), Chromium (24), Manganese (25), Iron (26), Cobalt (27) and Nickel (28). Though there is a mutual attraction between TITANIUM and either Nitrogen (7) or Oxygen (8), they may be best off avoiding one another.

INCOMPATIBILITY

Hydrogen (1), Beryllium (4), Carbon (6), Fluorine (9), Magnesium (12), Phosphorus (15), Sulfur (16), Chlorine (17), Potassium (19), Copper (29), Zinc (30), and Gallium (31) are not necessarily the best fits for TITANIUM. Additionally, although Nitrogen (7) and Oxygen (8) are highly attracted to TITANIUM, they may often be better off not engaging in relations.

ELEMENT BACKGROUND, GENERAL DISPOSITION QUALITIES

TITANIUM, is the 20th most common element in the Universe and the 9th most common on the Earth's surface. It is present in all living things. It is commonly found in many different minerals and is widely dispersed throughout the planet. TITANIUMs can generally be one of the more collected, stable, tolerant, and strong personality types. They value their character and integrity. They can also be some of the most giving and charitable people and enjoy sharing what they have with others. They tend to be outgoing and alluring. They can have a mystique about them too that is part of their allure. They are usually midland when it comes to making connections and relating to others. There are just as many personality types they can relate to versus those they do not. They are typically very good, some of the best, at processing and dealing with traditional day to day stress. However, they may not be the best at dealing with acute or elevated long-term stressors. TITANIUMs also strike a very good balance with their emotions and intuition versus thought and analysis. Nonetheless, they are ever so slightly more governed by their emotions and intuition versus thought and analysis when it comes to their assessment of others and situations. They are thinkers though; they enjoy contemplating many issues and appreciate thought provoking conversation and topics. TITANIUMs generally work very well with others and make great teammates/co-workers/partners. They can be initiators for many various things such as projects (work/home/group), taking the lead in certain situations, or just planning trips to name just some examples. They are typically very efficient and can handle much more than most with respect to workload and managing complex situations. They also tend to have a much higher tolerance level than most for a wide variety of both physical and mental demands. They are tough and resolute individuals. They typically are not a source of drama but can deal with and tolerate it better than most.

ADDITIONAL DEFINING QUALITIES, CHARACTERISTICS & INTERESTS

TITANIUM is a strong metal with a lustrous metallic-white appearance in its elemental state. It has the highest strength-to-weight ratio of all metals. It is very ductile; has a high melting point; is very resistant to corrosion and is relatively stable in acids and the air. Although, in time it slowly forms a protective layer when exposed to air. It also has a very high melting point. It is excellent with refracting light and absorbing ultraviolet rays. Though it is found in all animals, including humans, it is unknown for certain whether it is a necessity for animals - there is much to learn about it in this regard. However, it is believed that it stimulates carbohydrate production and encourages growth in plants. TITANIUMs are often content, well balanced and anchored persons; they have a very strong work ethic and work quite well with others. They can be dealt many hardships but will rarely complain. TITANIUMs can also be inward with their emotions; they may not express their negative emotions and feelings very often especially anything that can be perceived as complaining or that may cause concern for those around them. However, they are not reluctant to open up to those with their positive emotions and feelings. As a result, they are generally pleasant to be around. However, they can sometimes be skeptical of people. Especially if their life experiences have exposed them to repeated wrongdoings by people close to them; although this is true of most, TITANIUMs are more sensitive to this than most others. They are typically solid, altruistic, “salt of the earth”, reliable, and caring type people who are always willing to lend a hand to those around them. They can also be very protective of those that they care for. TITANIUMs typically care for the environment and enjoy travel, boats, planes/flying, the Sun, the water, hiking, the wilderness, electronics, pictures/photography, sports, jewelry, the arts, crafts, medicine/science, space/stars, cars, reading/writing, and various outdoor recreational activities.

23 THE DAY OF VANADIUM

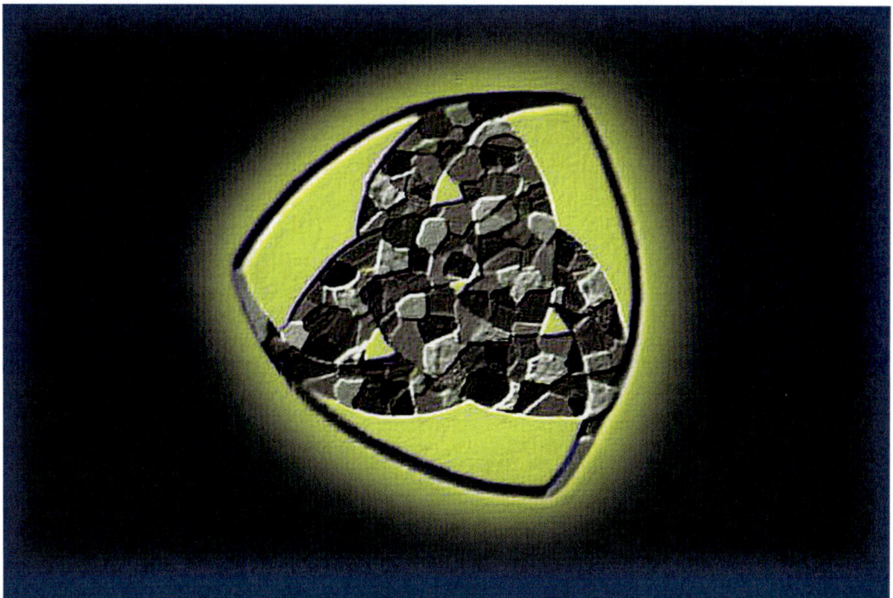

LIFE ELEMENT VANADIUM (V) - 23RD DAY BIRTH ELEMENT

CLASSIFICATION

VANADIUM is a solid metal at standard temperature and pressure and thus possesses Earth Element qualities. It is classified as a Transition Metal. It is very ductile, and is a good conductor of heat and electricity. It typically forms covalent bonds. Elements in the Transition Metal group are unique in the way that they can utilize more than one shell of electrons to combine with other elements. This allows them to form diverse and complex bonds.

COMPATIBILITY

VANADIUM personalities are most compatible with Life Elements Hydrogen (1), Oxygen (8), Silicon (14), Sulfur (16), Chlorine (17), and Calcium (20). They also can relate well with Sodium (11), Magnesium (12), Manganese (25), Copper (29), and Gallium (31). VANADIUMs also make mutually beneficial relations with Aluminum (13), Titanium (22) and Iron (26).

INCOMPATIBILITY

VANADIUM personalities are less likely to be compatible with Helium (2), Lithium (3), Beryllium (4), Boron (5), Carbon (6), Nitrogen (7), Fluorine (9), Neon (10), Phosphorus (15), Argon (18), Potassium (19), Scandium (21), Chromium (24), Cobalt (27), Nickel (28), and Zinc (30). Additionally, though there is an attraction, VANADIUMs and Oxygens (8) can sometimes result in virulent relations.

ELEMENT BACKGROUND, GENERAL DISPOSITION QUALITIES

VANADIUM is the 18th most common element in the Universe and also the 18th most common on the Earth's surface. It is found in very low levels in humans and is also present in all other living things. It exists in many different minerals throughout the world. VANADIUMs relate well with just as many Life Elements that they will have difficulties with. VANADIUMs share many characteristics with Titaniums (22) and these two Life Elements will get along well with one another. VANADIUMs are quite often very altruistic people who enjoy sharing what they have with others. They are also generally calm and tranquil people with an honorable and strong character. However, though they are calm and tranquil, they can also be pretty outgoing and vibrant. VANADIUMs can be known to over analyze; they can also find themselves in complicated situations, more so than most, with those which they have close relations. It is a good thing for them that they are also quite often very good when it comes to handling and dealing with stress and difficult situations. They are some of the best at holding up under prolonged stress. VANADIUMs are one of the most balanced Life Elements when it comes to their emotions/intuition and thought/analysis scale. They are governed equally by their emotions and intuition as opposed to thought and analysis. VANADIUMs typically work very well with others and make for excellent allies, colleagues, or companions. Also, similar to Titaniums (22), they are typically very efficient and can handle much more than most with respect to workload and stress; they also tend to have a much higher tolerance level than most for a wide variety of both physical and mental demands. They are typically not ones to stir up drama but they are better than most at tolerating it from other sources. They are often very thick skinned and don't allow much of anything to get to them.

ADDITIONAL DEFINING QUALITIES, CHARACTERISTICS & INTERESTS

VANADIUM is a medium-hard, malleable, ductile, and structurally strong metal with a bright white metal appearance in its elemental state. It has a relatively high melting point; it has good resistance to corrosion and is relatively stable in acids and the air. Though it is classified as an essential trace element for humans, its exact role is not known. Nonetheless it is suspected to play a part in the development of bones, teeth, cartilage, and preliminary studies show that it can control sugar levels in diabetics. It is found in marine algae, and also in very high concentrations in some other marine life, but its role in these species is somewhat a mystery too. VANADIUMs can be excellent listeners and be a very comforting source for those who are in times of need. They are typically very driven, stable and well balanced persons; they work well with others and can be unrelenting when there are goals to be reached. VANADIUMs like to get things done. As such they can make for great managers, leaders, role models, and motivators. VANADIUMs can be inward with their emotions in difficult, uncomfortable or confrontational times. This is a positive quality though as it results in providing them with the ability to remain calm and collected in contentious and stressful situations. They are typically reasonable, of strong character, and very reliable people who enjoy working hard and helping others.

They can also be protective of those that are dear to them. They may also at times exhibit outgoing, intense characteristics yet still maintain a "fun" demeanor. VANADIUMs typically enjoy working with their hands, travel, planes/flying, mechanics, the Sun, the water/the ocean, sweets, the outdoors, electronics, sports, the arts, crafts, aesthetics/anything with colors, medicine/science, and cars.

24 THE DAY OF CHROMIUM

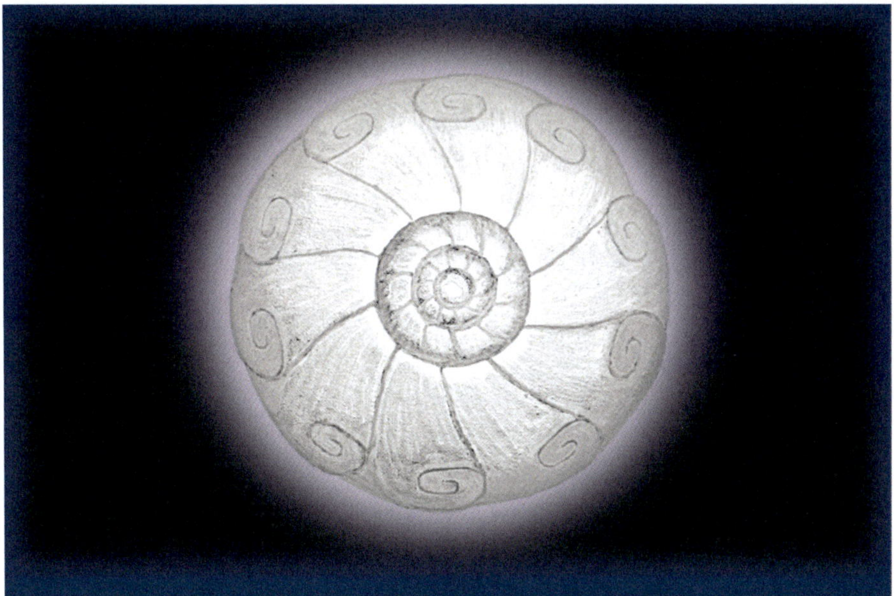

LIFE ELEMENT CHROMIUM (Cr) - 24TH DAY BIRTH ELEMENT

CLASSIFICATION

CHROMIUM is a solid metal at standard temperature and pressure and thus possesses Earth Element qualities. It is classified as a Transition Metal. It is hard and brittle; it is not a good conductor of electricity but it is an excellent conductor of heat. Elements in the Transition Metal group are unique in the way that they can utilize more than one shell of electrons to combine with other elements. This allows them to form diverse and complex bonds.

COMPATIBILITY

CHROMIUM personalities get along with a diverse mix of people. They are most compatible with Life Elements Oxygen (8) and Iron (26). They are also compatible with Hydrogen (1), Carbon (6), Magnesium (12), Silicon (14), Phosphorus (15), Sulfur (16), Potassium (19), and Iron (26). CHROMIUMs also make mutually beneficial relations with Aluminum (13), Titanium (22), Cobalt (27), Nickel (28), Copper (29), and Zinc (30). Though they may have an attraction and commonalities, CHROMIUMs should be aware that they can have destructive relations with some of their compatible Life Elements such as Oxygen (8) and Copper (29).

INCOMPATIBILITY

CHROMIUM personalities are less likely to be compatible with Helium (2), Lithium (3), Beryllium (4), Boron (5), Nitrogen (7), Fluorine (9), Neon (10), Sodium (11), Chlorine (17), Argon (18), Calcium (20), Scandium (21), Vanadium (23), Manganese (25), and Gallium (31). Although CHROMIUMs are very attracted to and can have many commonalities with either Oxygen (8), Potassium (19), or Copper (29) personalities, they sometimes have negative relations with these particular Life Elements.

ELEMENT BACKGROUND, GENERAL DISPOSITION QUALITIES

CHROMIUM is the 16th most common element in the Universe and the 20th most common on the Earth's surface. It is considered an essential trace element that is found in very low levels in humans. It is relatively common and present in many different compounds throughout the world. CHROMIUMs can relate well with many other Life Elements. However, there are also many that they will have some challenges with. Despite their challenges with particular Life Elements, they can have a vast and varying social circle. They are generally open-minded and tend to limit preconceived judgments. CHROMIUMs are often private, dependable, giving, and accommodating people. They are more often content, controlled, mild-mannered, and polished versus restless, wild, and rough. However, CHROMIUMs can sometimes exhibit another side in which they are not so mild-mannered or polished; they know how to let loose at times. They are usually very good when it comes to handling and dealing with normal day to day stress. However, they can crumble in traumatic or extremely high conflict situations. When it comes to their assessment of others and situations, CHROMIUMs are governed slightly more by their emotions and intuition over thought and analysis. CHROMIUMs often work very well under tough circumstances and conditions; and are tenacious at solving and working through problems and tasks. They also typically work very well with others and have very alluring, pleasing, and sometimes "smooth" personalities. They are visual creatures and appreciate beauty more than most. They are also rather affectionate and enjoy physical contact even if it is just a hug or innocent touching. They usually are not easily offended; and can be some of the most protective people of themselves and those they care for. They from time to time can be a source of drama as they can require a good deal of attention. They do not mind being the center of attention. CHROMIUMs can also be quite mindful of their appearance and how they are perceived by others despite not necessarily being over sensitive to criticism.

OTHER DEFINING QUALITIES, CHARACTERISTICS & INTERESTS

CHROMIUM is a hard, brittle metal with a lustrous steel-grey metal appearance. It is the hardest of all metals. It has a relatively high melting point; it has excellent resistance to corrosion and is relatively stable in many acids and the air at standard conditions. Though it is classified as an essential trace element for humans, its exact role is not known. However, it is suspected of playing a role in sugar and lipid metabolism; and in the signaling pathway of insulin. CHROMIUM is used to make numerous alloys such as stainless steel and is frequently used as a protective layer on metals/steels. It also combines to form numerous beautifully colored compounds and as such is used in many pigments and dyes. CHROMIUMs, though generally very interactive with others, can be somewhat guarded and closed with people they do not know. This is likely a result of their innate defense to protect themselves and those dear to them. However, once they have "vetted" newcomers they can be very inviting. They are generally motivated and hard working. CHROMIUMs can be all over the place with respect to expressing their emotions. That is, at times they can be open and even overbearing with discussing their thoughts, feelings, and emotions. At other times they can be closed and not willing to open up. They are typically reliable people who work hard and are helpful to those within their inner circle. CHROMIUMs typically are adventurous and mindful of their diet; they enjoy health and exercise, their appearance, electronics, sports, travel, the arts, music, medicine/science, history, foreign cultures, jewelry/clothing, cars/motorcycles, clothes, fashion; and they also appreciate aesthetics and colors.

25 THE DAY OF MANGANESE

LIFE ELEMENT MANGANESE (Mn) -
25TH DAY BIRTH ELEMENT

CLASSIFICATION

MANGANESE is a solid metal at standard temperature and pressure and thus possesses Earth Element qualities. It is classified as a Transition Metal. It is hard and very brittle. Oddly, given it's a metal, it is both a poor conductor of electricity and heat. Elements in the Transition Metal group are unique in the way that they can utilize more than one shell of electrons to combine with other elements. This allows them to form diverse and complex bonds.

COMPATIBILITY

MANGANESE personalities get along with a diverse mix of people. They are most compatible with Life Elements Carbon (6), Oxygen (8), Silicon (14), Titanium (22), and Iron (26). They are also compatible with Hydrogen (1), Lithium (3), Boron (5), Fluorine (9), Sodium (11), Phosphorus (15), Sulfur (16), Potassium (19), Calcium (20), Scandium (21), and Vanadium (23). MANGANESEs also make mutually beneficial relations with Magnesium (12), Aluminum (13), Nickel (28) and Copper (29).

INCOMPATIBILITY

MANGANESE personalities are less likely to be compatible with Helium (2), Beryllium (4), Nitrogen (7), Neon (10), Chlorine (17), Argon (18), Chromium (24), Cobalt (27), Zinc (30), and Gallium (31).

ELEMENT BACKGROUND, GENERAL DISPOSITION QUALITIES

MANGANESE, is the 17th most common element in the Universe and the 12th most common in the Earth's crust. It is an essential trace element for all living things. It is common and present in many different compounds throughout the world and is also found as a free element (by itself). MANGANESEs can relate well with many other Life Elements. They share many similar characteristics with Irons (26). MANGANESEs are often engaging, flexible, inviting, and sharing people. They can also be some of the most understanding people. They can be a great shoulder to lean on and make for wonderful friends, lovers, etc. because of their ability to make sincere efforts to both pick up on and understand others' feelings and emotions. They are typically genuine in the interests of others and even more so with those that they are close to. Those who know MANGANESEs the best will find they are an open book and keep very little to themselves. They are visual people that are inspired by observable stimuli (natural beauty, art, colors, etc.). They are typically below average to average when it comes to handling and dealing with stress. However, when they are supported by certain other Life Elements, they can handle stress and life's challenges with the best of them. MANGANESEs are perhaps one of the most balanced when it comes to their assessment of others and events as they are equally governed by their emotions and intuition versus thought and analysis. MANGANESEs often work well with others; have diverse interests and social circles. They genuinely enjoy the company of others; they generally seek out and require the company of others more than most. They are generally not a cause of drama and will try to avoid it. They typically do not handle acute drama, confrontation, and stressful situations very well; however, they are better at processing these things in small doses and over time. They can sometimes be mischaracterized in many ways, one should not judge them too quickly. They love their freedom and are at their best when those they have relations with are in tune with and allow them this. They will not respond well to those who try to "cage" them. Though it may not appear that they have it in them, MANGANESEs can go the extra mile especially when supported and encouraged by certain other Life Elements. They can be sensitive more than most, sentimental and even a bit corny; they can also be social butterfly types that enjoy and strive for harmony especially within themselves and between all people that they regularly interact with - they seek balance.

ADDITIONAL DEFINING QUALITIES, CHARACTERISTICS & INTERESTS

MANGANESE is hard, brittle with a silvery gray-white metal appearance. It has a low melting point (for a metal); it has some resistance to corrosion and is relatively stable in many acids and the air at standard conditions. However, it rusts like Iron (26) in moist air. It is an essential trace element for all animals and is essential to all living things. It plays varying roles in plants and animals. In plants it is responsible for producing Oxygen (8) through photosynthesis. In animals it plays pivotal roles in development, metabolism, and the antioxidant system (helps to remove harmful oxidants from the body). MANGANESE is used to make numerous alloys to increase the toughness, stiffness, wear resistance, hardness, and strength of metals. It also combines to form numerous beautifully colored compounds that are used in many pigments and dyes. MANGANESEs are usually pretty open and are able to let their guard down better than most. They can also be very tolerant and accepting of others, their opinions and lifestyles. This is very helpful to them because they thrive on friendships and relations with others; without having people in their lives that they can rely upon for friendship and intimacy they do not function at their optimal; this can lead to them being exposed to negative consequences such as depression, addictions, relating with others they should avoid, etc. Despite their requirement for interaction with others, MANGANESEs also require and enjoy alone time even though it may be only in small amounts. MANGANESEs can have a pretty balanced and healthy expression of their emotions. They are often very good at recognizing, assessing, and understanding situations/events/relations that are unstable and can be helpful with resolving such imbalances. MANGANESEs typically are somewhat adventurous and appreciate the environment/natural beauty, things that are aesthetically pleasing, art, intimacy, socializing, history, archeology, libations, construction, architecture, medicine/science, animals, and the water (the ocean, lakes, etc.). They also can have a tendency to want to learn and acquire knowledge of many different things. This could also include playing games that involve strategy or thought.

26 THE DAY OF IRON

LIFE ELEMENT IRON (Fe) - 26TH DAY BIRTH ELEMENT

CLASSIFICATION

IRON is a solid metal at standard temperature and pressure and thus possesses Earth Element qualities. It is classified as a Transition Metal. It is only one of three known magnetic elements. Elements in the Transition Metal group are unique in the way that they can utilize more than one shell of electrons to combine with other elements. This allows them to form diverse and complex bonds.

COMPATIBILITY

IRONs are most compatible with Life Elements Oxygen (8), and Silicon (14). IRONs can also make good relations with other Life Elements such as Hydrogen (1), Helium (2), Lithium (3), Beryllium (4), Carbon (6), Sodium (11), Magnesium (12), Aluminum (13), Phosphorus (15), Sulfur (16), Chlorine (17), Calcium (20), Scandium (21), Titanium (22), Vanadium (23), Chromium (24), Manganese (25), Copper (29), and Zinc (30). IRON is only one of three Life Elements that are ferromagnetic (magnetic) in its natural state. As such IRONs can have a high attraction for other IRONs, Cobalt (27), and Nickel (28).

INCOMPATIBILITY

IRON personalities are less likely to be compatible with Boron (5), Nitrogen (7), Fluorine (9), Neon (10), Argon (18), Potassium (19), and Gallium (31). Though they are very attracted and have many commonalities, IRONs can sometimes have turbulent relations with Oxygen (8). Though they can have very strong attractions to other IRONs, as well as Cobalt (27) and Nickel (28), they should be aware that they can also have some characteristics that are or can altogether be polar opposites and therefore be incompatible with these Life Elements.

ELEMENT BACKGROUND, GENERAL DISPOSITION QUALITIES

IRON is the 6th most common element in the Universe and the 4th most common in the Earth's crust; the Earth's core, however, consists mostly of IRON along with Nickel (28). When taking the Earth's core into account, IRON is by weight the most common Life Element on our planet and attributes to 1/3 of the weight of the Earth. It is the 13th most common element in the human body. It is essential for nearly all living things that we know of. It is one of the most widely utilized Life Elements, 2nd only to SILICON (14), and by far the most important of all metals to human beings. This is somewhat odd because it seems contradictory that it would be the most utilized metal given that it corrodes so easily; other elements are combined with it to prevent corrosion. IRONs share many similar characteristics with Manganese (25). IRONs are actually more often very sensitive, flexible, sharing, dependent, visual, and helpful people. However, at the same time and to a lesser degree, they can also have a side to them in which they can be very strong willed, rugged, "tough", inflexible and persistent. They are typically well rounded with many varied interests. They innately have more in common with many different Life Elements than those that they do not have much common ground with. As a result of their magnetic qualities, some can find them seductive and irresistible. They also can be the type of persons that stand up for others; and also for what they believe in. They are not ones to back down, especially when they have the support of others. On their own they are typically below average when it comes to handling and dealing with stress. However, with a support system from certain other Life Elements, they can be some of the best at processing and dealing with stress. This holds true in many other aspects of their lives more so than any other Life Element. Maintaining close and amicable relations with the right friends, family, work colleagues, and romantic interests is key for IRONs inner peace, sense of worth, will, drive, and ability to manage life's day to day troubles and stressors. Regarding their assessment of others and events in their lives, IRONs are very balanced with their emotions/intuition & thought/analysis. However, they are ever so slightly governed more by their emotions and intuition over thought and analysis. IRONs are often team players and strive in group settings; they can be excellent motivators when working with others, especially when tasks seem overwhelming and too daunting. They generally have a healthy expression of their emotions, feelings and thoughts. They can engage in and be the source of some drama but it is typically not too overbearing.

ADDITIONAL DEFINING QUALITIES, CHARACTERISTICS & INTERESTS

IRON is relatively soft, ductile with a silvery-white metal appearance. It has a high melting point; and is a moderately good conductor of heat and electricity. It easily corrodes; rusts in damp air; and reacts vigorously with some acids.

It is an essential trace element for nearly all plants and animals. It plays varying roles in plants and animals. For example, it is responsible for the red color of blood. It is an essential component of Hemoglobin, the substance that transports Oxygen (8) throughout the body. In plants, IRON plays a role in the creation of chlorophyll. IRONs have much in common with Manganese (25). IRON is alloyed with numerous other elements to reduce corrosion and increase the toughness, stiffness, wear resistance, hardness and strength; it is the most widely used metal by far and the basis of all steel products. It also combines to form numerous beautifully colored compounds that are used in many pigments and dyes. IRONs are usually pretty open and readily able to let their guard down when engaging with others. Similar to Manganese's (25), IRONs can also be very tolerant and accepting of others, their opinions and lifestyles. This is convenient for them because they flourish on friendships and relations with others; without having people in their lives that they can rely upon for friendship, support and intimacy they do not function well. Deficiencies in this area can lead to them being exposed to adverse ramifications such as depression, addictions, or having relations with others they should avoid. IRONs can have a pretty balanced and healthy expression of their emotions. They are often "softies"; this does not mean they are pushovers. Rather it is that they have a sensitive and caring disposition; this more often is not apparent to most. IRONs can have a sixth sense when it comes to recognizing imbalances with relations they have between themselves and others; and also at recognizing imbalances with relations between others that they are observers of. Thus, they can be excellent counselors, mediators and friends. IRONs generally have a curiosity for "how things work" and have interests with electronics, medicine, the environment, games, working with their hands, a wide variety of recreational activities, spirituality, architecture, construction, engineering, mechanics, interior design/decor, reading, writing, history, cultures, jewelry, culinary/cooking, technology, music, dining and food, boats, cars, road trips, travel, and the outdoors.

27 THE DAY OF COBALT

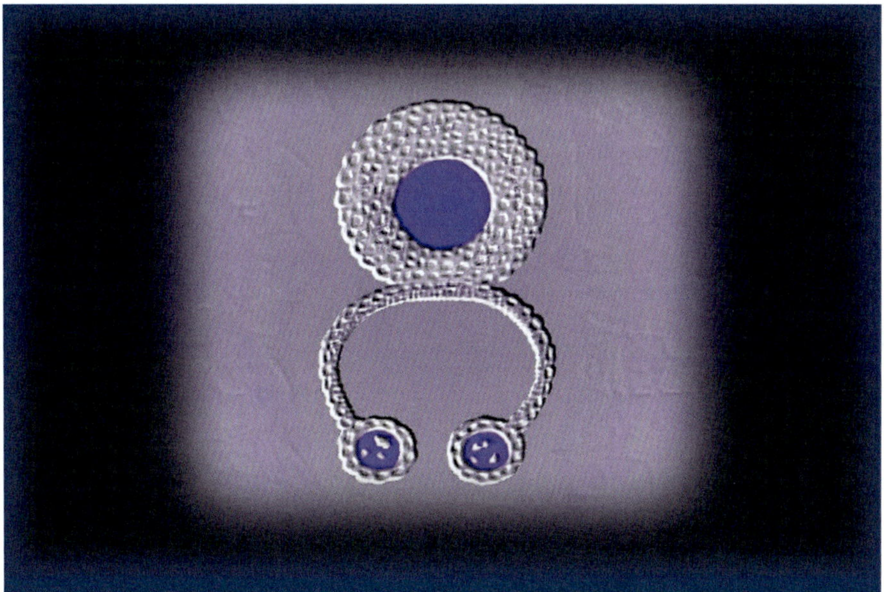

LIFE ELEMENT COBALT (Co) - 27TH DAY BIRTH ELEMENT

CLASSIFICATION

COBALT is a solid metal at standard temperature and pressure and thus possesses Earth Element qualities. It is classified as a Transition Metal. It easily forms stable compounds with neutral molecules. It is only one of three known magnetic elements. Elements in the Transition Metal group are unique in the way that they can utilize more than one shell of electrons to combine with other elements. This allows them to form diverse and complex bonds.

COMPATIBILITY

COBALT personalities are more likely to be compatible with Life Elements Carbon (6), Oxygen (8), Sulfur (16), Calcium (20), and Copper (29). They also can have mutually beneficial relations with Aluminum (13), and Titanium (22), Chromium (24), and Zinc (30). COBALT is only one of three Life Elements that are ferromagnetic (magnetic) in its natural state. As such, COBALTs can have a high attraction to other COBALTs (27), Iron (26) and Nickel (28) personalities. COBALT personalities are unique in that they can often get along well with nearly everyone despite having little or no connection.

INCOMPATIBILITY

As mentioned, COBALT is rather stable, and it interacts relatively calmly with other elements. However, Hydrogen (1), Helium (2), Lithium (3), Beryllium (4), Boron (5), Nitrogen (7), Fluorine (9), Neon (10), Sodium (11), Magnesium (12), Silicon (14), Phosphorus (15), Chlorine (17), Argon (18), Potassium (19), Scandium (21), Vanadium (23), Manganese (25), and Gallium (31) may not be the best fits for COBALT. COBALT personalities are quite unique though in the way that they can get along with these Life Elements as well as any other Life Element despite not necessarily having much in common or having an attraction of some sort. Though they can have very strong attractions to other COBALTs as well as IRONs (26) and Nickel (28), they can also have some characteristics that are or may altogether be polar opposites and therefore be incompatible with these Life Elements.

ELEMENT BACKGROUND, GENERAL DISPOSITION QUALITIES

Cobalt is the 19th most common Life Element in the Universe. It is found in low levels on Earth and in the human body. COBALT is a trace element though it is relatively common and widely dispersed throughout the planet and it is found in both the water (oceans, rivers, and lakes) and the Earth's crust. It is present in all animals and plants. Cobalt is a relatively stable element that has very diverse characteristics. COBALTs tend to be very content, stable, and possess a wide range of interests. Most people born on the 27th day find it easy to get along with most, even those they may not agree with or even admire. This is true even though the number of Life Elements they intrinsically relate well with is far outnumbered by those they do not have an innate connection with; this is a quality unique to COBALTs. COBALTs can possess "magnetic" personalities - many are drawn to them; or they can be seen as someone to be "desired". They are typically very likeable and can have very attractive personalities; this despite the fact that they can be somewhat reserved and conservative. However, they often may also reserve judgment of others, situations, and issues which can be a very positive quality for them. As a result of their ability to tolerate even those they may not appreciate, they also tend to be very good listeners, take their time to assess others; and are accepting and understanding of others. They are also accepting of the general circumstances that they find themselves. That is, they are generally content with their position and place in life. However, this does not mean they lack motivation by any means. They are above average when it comes to dealing with stress, but some things will get to them slowly over time. They are generally thoughtful of others and of the big picture; they often recognize and perceive their surroundings and the world as a vast complex "interactive system" versus a simplistic self-centered small world that only involves them and those around and close to them. They are governed more by thought and analysis over emotion and intuition. They can also be protective of those for whom they care about. Drama is not their style but they can tolerate it pretty well.

ADDITIONAL DEFINING QUALITIES, CHARACTERISTICS & INTERESTS

COBALT is silver-white with a hint of blue solid metal in its elemental state. COBALT is very resistant to heat. It has the highest known temperature at which it retains its magnetic properties. It is essential to many life forms and is present in plants and animals. It is a fundamental component of Vitamin B_{12} which is essential to humans and most other animals. It plays a key role in the normal functioning of the brain and nervous system; and is involved in the metabolism of every cell of the body. COBALT also forms more known complex ions than any other Life Element metal. It is used in many ceramics. COBALT is tough, ductile, malleable and a conductor of heat and electricity. COBALTs are very good at maintaining their innate stable disposition under extreme pressure; they are very versatile and make for excellent workmates/teammates/etc., and often possess excellent leadership skills especially in difficult and complex situations. They are also very good at diffusing combative situations and make very good liaisons. COBALTs can be very deep and complex thinkers, highly contemplative and introspective. They are sensitive, but do not necessarily outwardly display this as they are not the most expressive of their feelings in general. They typically are content with who they are and their position. They value family, friendships and very much require companionship as most of us do. COBALTs can be inward with their emotions; but in times of distress they can be very open and calm toward others. Many COBALT personalities are lucky in that they can be very friendly and this trait comes naturally to them. COBALTs typically love the outdoors and water; don't mind getting dirty, and are "steady as you go" type people. They are lovers of life who enjoy many different topics such as: the arts, fashion, politics, sports, medicine, history, cultures, archeology, science, aesthetics, cars, travel, animals and people.

28 THE DAY OF NICKEL

LIFE ELEMENT NICKEL (Ni) - 28TH DAY BIRTH ELEMENT

CLASSIFICATION

NICKEL is a solid metal at standard temperature and pressure and therefore possesses Earth Element qualities. It is classified as a Transition Metal. It is only one of three known magnetic elements. Elements in the Transition Metal group are unique in the way that they can utilize more than one shell of electrons to combine with other elements. This allows them to form diverse and complex bonds.

COMPATIBILITY

NICKELs are most compatible with Life Elements Sulfur (16) and Iron (26). They are also compatible with Hydrogen (1), Carbon (6), Nitrogen (7), Oxygen (8), Magnesium (12), and Phosphorus (15). NICKELs also make mutually beneficial relations with other Life Elements such as Beryllium (4), Titanium (22), Chromium (24), Manganese (25), Copper (29), and Zinc (30). NICKEL is only one of three Life Elements that are ferromagnetic (magnetic) in its natural state. As such NICKELs can have a high attraction for other NICKELs, Irons (26) and Cobalts (27).

INCOMPATIBILITY

NICKEL personalities are less likely to be compatible with Helium (2), Lithium (3), Boron (5), Fluorine (9), Neon (10), Sodium (11), Aluminum (13), Silicon (14), Chlorine (17), Argon (18), Potassium (19), Calcium (20), Scandium (21), Vanadium (23), and Gallium (31). Though they can have very strong attractions to other NICKELs, as well as Irons (26) and Cobalts (27), NICKELs should be aware that they can also have some characteristics that are or may altogether be polar opposites and therefore be incompatible with these Life Elements.

ELEMENT BACKGROUND, GENERAL DISPOSITION QUALITIES

NICKEL is the 13th most common element in the Universe and is very common on Earth. It makes up a large portion of the Earth's core along with Iron (26). It is a trace element in the human body. It is a relatively unreactive element and does not readily combine with Oxygen (8) or water nor does it dissolve in most acids. It is considered non-corrosive and relatively stable. NICKELs are in the middle when it comes to having innate commonalities and connections with others; they pretty much have an even amount of Life Elements that they are compatible with versus those they are not. NICKELs are generally very tolerant of and considerate of others. They are also typically stable and reasonable people that many are drawn to. Though they are commonly accommodating to others, they can also be tough (not stubborn, but stern) and not easily swayed. They are thick-skinned in the manner that they are not easily offended and don't let things get to them too readily. They take most of life's difficulties with a grain of salt. When life gives them lemons, they make lemonade. They are also resolute in certain aspects of their lives. In particular, they prefer to steer their own ship and though they are open to other points of view and suggestions, they do not like to be oppressed in any facet of their lives; be it relations with friends, lovers, family or workmates/bosses, etc. They like freedom and do not like to be caged. They are typically very good, some of the best in fact when it comes to handling and dealing with stress. NICKELs are governed by thought and analysis over emotion and intuition. NICKELs will take on all they can handle and more while working very hard with little or no complaints. They can surprise many with how much effort and determination they are capable of; they are quite resilient. They thrive in group settings as this brings out these characteristics even more. They can also often be the hard working quiet person that offers simple solid solutions. They typically are not the source of drama. However, they have a very high tolerance for it, and thus they can get along with those that are melodramatic even to the extreme. NICKELs can be very protective of those that they care about.

ADDITIONAL DEFINING QUALITIES, CHARACTERISTICS & INTERESTS

NICKEL is hard, ductile, and malleable with a lustrous silvery-white metal appearance with a minor gold tinge. It has a relatively high melting point; and is a moderately good conductor of heat and electricity. It is highly resistive to corrosion; and reacts rather mildly with acids. NICKEL is essential for the proper function of some enzymes in plants and animals. It is also involved in the transmission of genetic codes for DNA and RNA. NICKEL is alloyed with numerous other elements to increase the toughness, stiffness, wear resistance, hardness and strength. It is widely used for alloying with and plating other metals. It also combines to form numerous colored compounds that are used as coloring agents in paints. NICKELs tend to lean toward the shy and reserved side but nonetheless can be engaging and open towards others. NICKELs are generally slightly reserved when it comes to expressing their emotions. However, in private, with those they are most familiar with, they tend to have a very healthy expression of their emotions. The more you get to know them, the more likely than not you know where you stand with NICKELs as they are typically straight forward individuals. Although they can be more direct with those who are close to them, at the same time they carry themselves with a level of tact to both those close to them as well as others in general; in other words, they can be forward but are perceptive and courteous. They appreciate efficiency and also tend to be logical and pragmatic. They can be very good at understanding and/or explaining complex and detailed issues. NICKELs will often have interests with medicine, space, brain teasers/games, their appearance, staying fit, the water/ocean/lakes, science, electronics, technology, architecture, construction, music, the arts, jewelry, culinary/cooking, dining and food, military, history, money, travel and foreign cultures. They are also visual and appreciate the many aesthetic beauties in nature.

29 THE DAY OF COPPER

LIFE ELEMENT COPPER (Cu) - 29TH DAY BIRTH ELEMENT

CLASSIFICATION

COPPER is a solid metal at standard temperature and pressure and thus possesses Earth Element qualities. It is classified as a Transition Metal. It differentiates itself from all other elements in that it has both an extremely high electrical and thermal conductivity. Elements in the Transition Metal group are unique in the way that they can utilize more than one shell of electrons to combine with other elements. This allows them to form diverse and complex bonds.

COMPATIBILITY

COPPERs are most compatible with Life Elements Carbon (6), Oxygen (8), Sulfur (16), Iron (26) and fellow COPPERs. They are also compatible with Hydrogen (1), Sodium (11), Phosphorus (15), Chlorine (17), Potassium (19), Cobalt (27), and Gallium (31). COPPERs also make mutually beneficial relations with other Life Elements such as Beryllium (4), Aluminum (13), Silicon (14), Calcium (20), Vanadium (23), Chromium (24), Manganese (25), Iron (26), Nickel (28), and Zinc (30). Although COPPER and Chromium (24) have commonalities and an attraction for one another, these two Life Elements can have tumultuous relations.

INCOMPATIBILITY

COPPER personalities are less likely to be compatible with Helium (2), Lithium (3), Boron (5), Nitrogen (7), Fluorine (9), Neon (10), Magnesium (12), Argon (18), Scandium (21), and Titanium (22). Although Copper is attracted to and can have many commonalities with Chromium (24) personalities, they sometimes have negative relations

ELEMENT BACKGROUND, GENERAL DISPOSITION QUALITIES

COPPER is the 27th most common element in the Universe and the 24th most common on the Earth's surface. It is an essential trace element in the human body as well as in all other living organisms including plants and animals. It is found on Earth in its native state as well as in a vast variety of compounds. It does not readily combine with Oxygen (8) or water; it resists the weathering of the atmosphere and seawater. It is not soluble in acids; and resists corrosion from most acids. Therefore, it is considered to be highly resistant to corrosion and is relatively stable. COPPERs are generally open persons; they are also commonly rather tolerant of others views and lifestyles. Though they do get along with and have much in common with most other people, they can also find that they have challenges relating to certain other personality types. They generally tend to be content, warm and stable persons who sometimes have unique perspectives and lifestyles. They are curious and will not always conform to the norm. They also tend to be minimalists who value their possessions from a sentimental point of view rather than from a monetary or status perspective. This does not mean they do not appreciate wealth. They typically do not let too many things bother them or get under their skin. They value their "freedom" more than most and can exhibit a wanderlust side. This is not to say that they are lost souls but rather: they are at their best when they are allowed to express their uniqueness. They can be creative in varying capacities. They may also have a deep connection with experiencing foreign lands and learning other cultures and their history. They can be protective of both their possessions and those who are near to them. When in tune, they process stress quite well; they are at the high end when it comes to handling and dealing with stress. This is mainly due to their ability to accept things for what they are, compartmentalizing, and that they prefer to avoid stressors altogether if they can. COPPERs are governed by thought and analysis more than emotion and intuition. COPPERs are also unique in the way that many can be very perceptive of the "energy" that others emanate, more so than any other of the Life Elements. COPPERs work moderately well with some and better with others. They typically are not a source of drama. However, they do have a decent tolerance for it. COPPERs can often be the type that "beat to the sound of their own drum". COPPERs may not necessarily present such, but they can endure beyond any expectation - they are survivors.

ADDITIONAL DEFINING QUALITIES, CHARACTERISTICS & INTERESTS

COPPER is soft, ductile, and malleable with a reddish-orange metal appearance. It has a moderate melting point; and is an extremely good conductor of heat and electricity. It is resistive to corrosion; and it either does not react or reacts rather mildly with most acids. It is essential for all life. It is the key component of redox enzymes (transfer of electrons) and of haemocyanin (a protein containing copper, responsible for transporting oxygen in the blood plasma of arthropods and mollusks). Copper proteins have diverse roles in biological electron transport and oxygen transport. COPPER is also essential for the proper function of some enzymes in plants and animals. COPPER is still the preferred metal for wiring, and thus is very widely utilized in homes, buildings, electronics, cars, machines and nearly everything that is powered by electricity. COPPER is alloyed with numerous other elements for its aesthetic appearance, resistance to corrosion, and more and more for sanitary purposes. COPPER has excellent anti-microbial qualities. That is, COPPER resists bacteria, fungus, and viruses and kills or prevents growth of these on its surface. It is also used for plating other metals. It also combines to form numerous colored compounds that are used as coloring agents and pigments. COPPERs require and enjoy alone time very much; they can be very content alone or in the company of very few for extended periods but they also will seek a balance as they have the need for experiencing life beyond their inner circle. COPPERs are generally pretty open and balanced when it comes to expressing their emotions. They can make for excellent facilitators in many aspects as they enjoy helping others make things happen. They are also often very good listeners and can be very helpful to others in this regard. They are usually efficient with their time and finances; and tend to have a good balance between spirituality, logic, and pragmatism. Balance is a common theme for COPPERs as they seek it in all facets of their lives. COPPERs love nature and the outdoors. They will commonly be “earthy” and appreciate aesthetics and the multitude of visible beauties that Nature offers. They will often have interests with medicine, personal hygiene, cleanliness, their appearance and aesthetics in general, photography, animals, plants, the outdoors, camping, the water, science, the environment, mechanics, technology, architecture, construction, how things work, all of the arts - and especially visual, reading, jewelry, culinary/cooking, dining, money, travel, exploration, music, fashion, home décor, history, archeology, and anthropology.

30 THE DAY OF ZINC

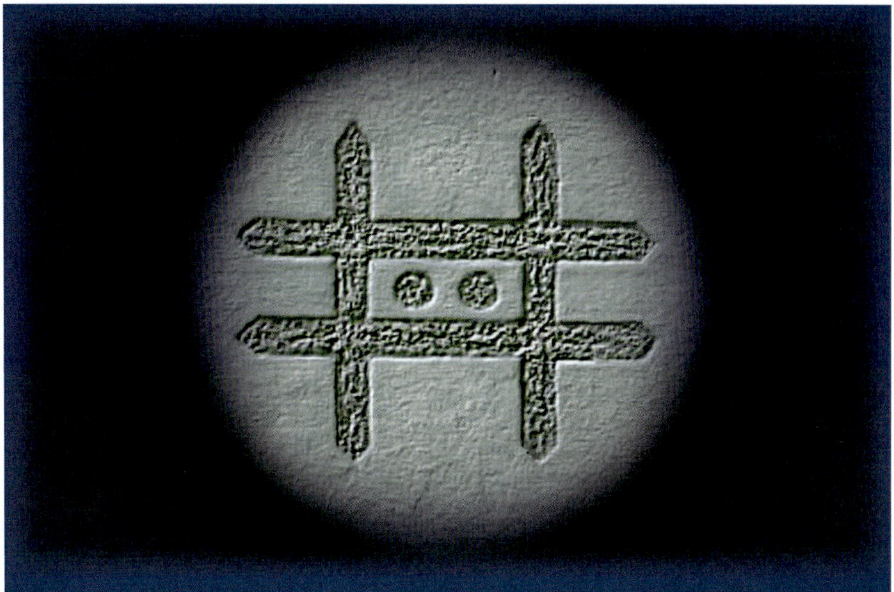

LIFE ELEMENT ZINC (Zn) - 30TH DAY BIRTH ELEMENT

CLASSIFICATION

ZINC is a solid metal at standard temperature and pressure and thus possesses Earth Element qualities. It is classified as a Transition Metal. Elements in the Transition Metal group are unique in the way that they can utilize more than one shell of electrons to combine with other elements. This allows them to form diverse and complex bonds.

COMPATIBILITY

ZINCs are most compatible with Life Element Sulfur (16). They also get along well with Carbon (6) and Silicon (14). Also, Oxygen (8), Phosphorus (15), Chlorine (17), Calcium (20), Chromium (24), Copper (29), and Gallium (31) are compatible with ZINC. ZINCs also make mutually beneficial relations with other Life Elements such as Aluminum (13), Iron (26), Cobalt (27), and Nickel (28).

INCOMPATIBILITY

ZINC personalities are less likely to be compatible with Hydrogen (1), Helium (2), Lithium (3), Beryllium (4), Boron (5), Nitrogen (7), Fluorine (9), Helium (10), Sodium (11), Magnesium (12), Argon (18), Potassium (19), Scandium (21), Titanium (22), Vanadium (23), and Manganese (25).

ELEMENT BACKGROUND, GENERAL DISPOSITION QUALITIES

ZINC is the 25th most common element in the Universe and the 23rd most common on Earth. It is an essential mineral in the human body, and all living organisms including plants and animals. It does not react with Oxygen (8) in dry air. However, it tarnishes rather quickly upon exposure to moist air - this provides it with a protective layer which substantially slows further corrosion; it resists the weathering of the atmosphere. ZINCs generally have a wide range of interests and are curious by nature, generous, unselfish, protective, loyal, engaging, and very nurturing persons. They are also typically open to others opinions, though they can sometimes come across to the contrary by appearing stiff and rigid. They get along with and have much in common with many, but they equally will find that they have challenges relating with many others as well. They can be very guarded at first when meeting and getting to know others but will open up over time. They very much enjoy the company of others and prefer to spend most of their time and activities with others. They place a lot of value on their relationships (neighbors, workmates, friends, family, lovers, etc.) and can be very protective of them. They can be the "helicopter" type mom; and they often worry about too many things, many of which are beyond their control. With that said though, they are generally pretty good at dealing with stress. Also, though they are very involved with, and have a lot of emotional capital invested in their relationships, they are more governed by thought and analysis over emotion and intuition. They can also be very expressive of their feelings and emotions but not necessarily in a bad way as much of their thoughts are typically well filtered; they are thoughtful and calculated (more often in a good way). ZINCs will more often make for excellent workmates, partners, and colleagues. However, because of their protective nature and desire to "regulate"/ shape the way things are to an extent, they can be overbearing at times to some and generally all the time to some others. They can also be wonderful project managers and managers in general so long as they leave their "helicopter" at home because this can lead to them being micromanagers. ZINCs need to pay close attention to this "helicopter" characteristic and regulate it. It can be present in many areas of their personalities and thus cause disruptions to various parts of their lives. They have the ability to sense and perceive the energies of others especially of those they are more familiar with. They can be knowledge seekers; take and endure a lot; and can be resilient and exhibit a high endurance in multiple areas of their lives. They can also be very altruistic people and will sacrifice their time, effort, and own advancement for the sake of certain other individuals. ZINCs can be a source of drama; however, they also have a lot of patience for it too. ZINCs can be the type to sacrifice of themselves whatever it takes for the development/betterment or well-being of those they care for. ZINCs are definitely good to have on your side.

ADDITIONAL DEFINING QUALITIES, CHARACTERISTICS & INTERESTS

ZINC is hard and brittle with a bluish-silver, lustrous metal appearance. It has a low melting point for a metal. It is a fair conductor of electricity and has a high thermal conductivity. It is resistive to corrosion; it reacts very slowly with weak acids. ZINC plays many roles in the biology of plants, animals, and microorganisms. It is the primary component of hundreds of enzymes; plays a pivotal role in the development of many chemicals in animals. The protein hormone insulin contains ZINC, and there is much research being conducted surrounding diabetes and the known depletion of ZINC in cells that produce insulin. ZINC is also used to build molecules of DNA. ZINC is also essential for the proper growth, reproduction, and function of plants and animals. As an amazing example of how important and sensitive we are to the elements, the recommended daily intake of ZINC for an adult is around 10 mg/day or 0.01 grams. That is 1/100th the weight of a dollar bill! Animals lacking the required amounts of ZINC (say they are lacking just 0.005 grams or 1/200th the weight of a dollar bill per day) need 50% more food to gain the same weight of an animal with sufficient levels of ZINC. ZINC is alloyed with some other elements for its aesthetic appearance and resistance to corrosion. It is also widely used for plating metals, particularly Iron (26) and steels, to prevent corrosion. ZINCs require and enjoy time with others very much; they will likely become restless when alone for extended periods. ZINCs are generally pretty open and balanced when it comes to expressing their emotions. They enjoy helping others; are good listeners; can be very protective; and often offer sensible and sound advice. ZINCs also like to maintain balance in many facets of their lives. They may often make conscious efforts to regulate their daily behaviors to ensure they preserve this balance. They can be the source of many beginnings whether it be just making introductions; match making, forming clubs, creating anything, or planning and executing ideas/projects, etc. ZINCs are drawn to people, animals, and Nature in general. They will often have interests with photography, electronics, medicine, animals, plants/flowers, personal hygiene, fashion/clothes, the water, science, art, music, organizations, socializing, technology, architecture, construction, mechanics, and how things work in general, reading and or writing, exploring, travel, fashion, home décor, cars, cultures, anthropology, TV/movies and history.

31 THE DAY OF GALLIUM

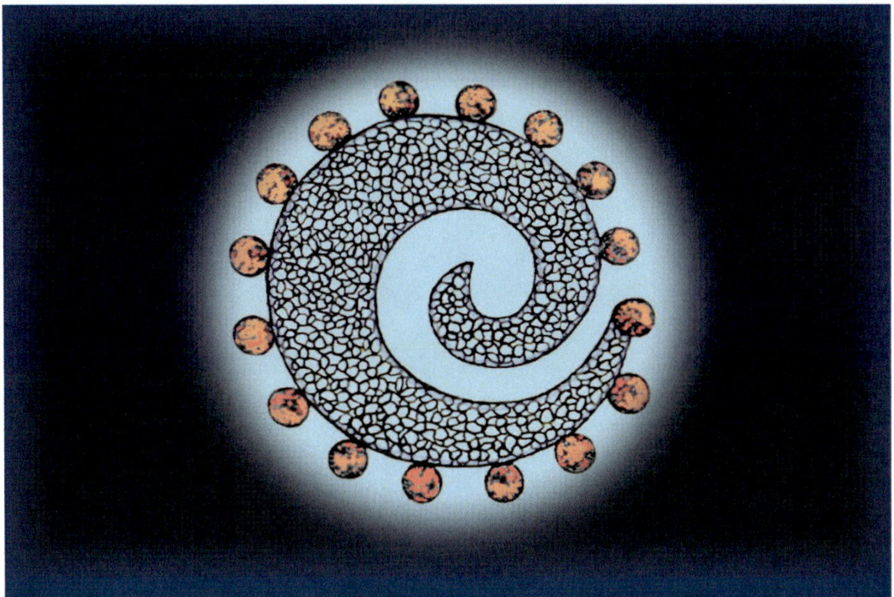

LIFE ELEMENT GALLIUM (Ga) - 31ST DAY BIRTH ELEMENT

CLASSIFICATION

GALLIUM is a solid metal at standard temperature and pressure and thus possesses Earth Element qualities. However, given that it turns to liquid when held in the hand, it also holds Water Element qualities as well. It is classified as an Other Metal because of its many non-metal like characteristics. Elements in the Other Metal group only use their outer shell electrons to combine with other elements.

COMPATIBILITY

GALLIUMs are most compatible with Life Elements Carbon (6), Oxygen (8), Aluminum (13), Sulfur (16), Argon (18), Vanadium (23), Copper (29), and Zinc (30).

INCOMPATIBILITY

GALLIUM personalities are less likely to be compatible with Hydrogen (1), Helium (2), Lithium (3), Beryllium (4), Boron (5), Nitrogen (7), Fluorine (9), Helium (10), Sodium (11), Magnesium (12), Silicon (14), Phosphorus (15), Chlorine (17), Potassium (19), Calcium (20), Scandium (21), Titanium (22), Chromium (24), Manganese (25), Iron (26), Cobalt (27), and Nickel (28).

ELEMENT BACKGROUND, GENERAL DISPOSITION QUALITIES

GALLIUM is not a very common element in the Universe or on Earth. Though it is not one of the more abundant Life Elements, it is widely dispersed throughout the surface of the Earth. It is found in the human body, but only in very minute amounts. GALLIUM does not appear to play a role in any living organisms including plants and animals. It is stable in air and water because it forms a protective oxide layer that further prevents it from reacting with Oxygen (8) or water. GALLIUM is fairly reactive in relation to other elements. It reacts with both acids and bases. It is unique in that it is solid at room temperature but will turn to liquid if taken outside on a hot day or held in the hand (its melting point is just over 85 degrees Fahrenheit). It also has the widest temperature range of all elements in its liquid form. GALLIUMs tend to be givers; they are also generally curious, reflective, self-aware, spontaneous, and adventurous people. They also typically find it easy to adapt in many different situations and scenarios; they generally handle change with ease. They are also capable of viewing situations, issues, and problems from multiple perspectives. This provides them with both the ability to keep an open mind and possess great problem solving capabilities.

They are facilitators and enjoy sharing their energies to help others. They can be slightly guarded at first when meeting and getting to know others. They enjoy others and social interaction but it is not necessarily a top priority for them and they can be content on their own for extended periods of time. They will struggle relating to most other Life Elements. GALLIUMs generally lie at the lower end of the middle when it comes to dealing with stress. They are more governed by thought and analysis over emotion and intuition. GALLIUMs are less likely to be a source of drama and do not handle it well. GALLIUMs can be very good thinkers on their feet and in general.

ADDITIONAL DEFINING QUALITIES, CHARACTERISTICS & INTERESTS

GALLIUM has a silvery metal appearance. It has a very low melting point; in fact, it liquefies when held in the hand but will refreeze when removed from the hand. It has the greatest span of temperature in its liquid state. It is only one of two Life Elements, Silicon (14) being the other, which expands in its solid state (expands upon freezing). Certain GALLIUM compounds have shown very promising results for potential treatments for numerous types of cancer, inflammation, and infectious diseases. It is also said to stimulate the metabolism. It is a low conductor of electricity and thermal conductivity. GALLIUM is used in electronic circuit boards, converting light to energy, LED lights and mirrors. GALLIUMs very much enjoy social interaction but their options will be limited as they struggle to make connections with many other Life Elements. However, they do have many characteristics that help them get along well with others at least on a superficial level. They also have other attributes that provide them with a curiosity for and judgment skills of others. They are observers of people and are always assessing the actions and motivations of individuals. GALLIUMs often have a balanced and healthy expression of their feelings and emotions. They typically have a connection with science and or medicine. They often have unique perspectives to offer which can be beneficial in friendships and work settings. This also can give rise to a creative advantage. They can be very aware of and place much emphasis on how they present themselves. This can be seen in their physical appearance, demeanor, and their words. Though they may sometimes come across to the contrary, they can actually be very sensitive people. GALLIUMs will often have interests in medicine, learning, the Sun (being outside), science, technology, computers, electronics, reading and or writing, creativity, exercise, travel, nightlife, the environment, fashion, and staying physically active.

SOURCES AND CREDITS

Photos from NASA JPL are used in full or have been altered/changed to create the artwork for each of the following:

The cover page, Hydrogen, Helium, Lithium, Boron, Oxygen, Neon, Argon, and Titanium.

The following websites and articles were used throughout this publication as sources of information:

http://hyperphysics.phy-astr.gsu.edu/hbase/tables/ttab.html#c1
http://en.wikipedia.org/wiki/Abundance_of_the_chemical_elements
http://periodictable.com/Properties/A/UniverseAbundance.html
http://chemwiki.ucdavis.edu/Inorganic_Chemistry/Case_Studies/
Case_Study%3A_The_Natural_Abundance_of_Elements
http://antoine.frostburg.edu/chem/senese/101/periodic/faq/what-element-is-most-abundant.shtml
http://chemwiki.ucdavis.edu/Inorganic_Chemistry/Descriptive_Chemistry/d-
Block_Elements/Group_9%3A_Transition_Metals/Chemistry_of_NICKEL
http://en.wikipe facts/periodic_table/NICKEL_facts/203/
https://www.webelements.com/NICKEL/
http://www.chemicool.com/elements/NICKEL.html
http://www.rsc.org/periodic-table/element/27/NICKEL
http://chemwiki.ucdavis.edu/Inorganic_Chemistry/Descriptive_Chemistry/Main_
Group_Reactions/Reactions_of_Main_Group_Elements_with_Oxygen
http://education.jlab.org/glossary/abund_ele.html
http://www.britannica.com/science/hydrogen
http://www.rsc.org/periodic-table/element/1/hydrogen
http://www.encyclopedia.com/topic/hydrogen.aspx
http://www.chemistryexplained.com/Ge-Hy/Hydrogen.html
https://www.webelements.com/hydrogen/
http://www.softschools.com/facts/periodic_table/hydrogen_facts/177/
http://en.wikipedia.org/wiki/Hydrogen
https://www.boundless.com/chemistry/textbooks/boundless-chemistry-textbook/
nonmetallic-elements-21/hydrogen-148/properties-of-hydrogen-571-3498/
http://www.chemicool.com/elements/hydrogen.html
http://chemwiki.ucdavis.edu/Inorganic_Chemistry/Descriptive_Chemistry/s-Block_Elements/
Group__2_Elements%3A_The_Alkaline_Earth_Metals/Chemistry_of_Magnesium
http://en.wikipedia.org/wiki/Magnesium
http://education.jlab.org/itselemental/ele012.html
http://www.rsc.org/periodic-table/element/12/magnesium
http://chemwiki.ucdavis.edu/Inorganic_Chemistry/Descriptive_Chemistry/
p-Block_Elements/Group_17%3A_The_Halogens
http://encyclopedia2.thefreedictionary.com/Magnesium
http://www.britannica.com/science/magnesium
http://periodic.lanl.gov/12.shtml
https://www.webelements.com/magnesium/
http://www.livescience.com/28862-magnesium.html
http://www.webelements.com/helium/chemistry.html
https://www.chem.wisc.edu/deptfiles/genchem/sstutorial/Text5/Tx54/tx54.html

http://www.rsc.org/periodic-table/element/2/helium
http://en.wikipedia.org/wiki/Helium
http://www.chemicalelements.com/groups/alkali.html
http://www.chemicalelements.com/groups/alkaline.html
https://www.boundless.com/chemistry/textbooks/boundless-chemistry-textbook/metals-20/metals-145/the-alkaline-earth-metals-565-6854/
http://www.britannica.com/EBchecked/topic/15638/alkaline-earth-metal/80892/Physical-and-chemical-behaviour
http://www.gcsescience.com/pt14.htm
http://chemed.chem.purdue.edu/genchem/topicreview/bp/ch12/trans.php
http://www.chemicalelements.com/groups/transition.html
http://www.chemicalelements.com/groups/nonmetals.html
http://www.chemicalelements.com/groups/noblegases.html
http://www.chemicalelements.com/groups/othermetals.html
http://www.chemicalelements.com/groups/halogens.html
http://chemwiki.ucdavis.edu/Inorganic_Chemistry/Descriptive_Chemistry/Elements_Organized_by_Block/2_p-Block_Elements/Group_17%3A_The_Halogens/Chemistry_of_Fluorine
https://en.wikipedia.org/wiki/Fluorine
http://www.webelements.com/fluorine/
http://www.encyclopedia.com/topic/fluorine.aspx
http://www.britannica.com/science/fluorine
http://www.livescience.com/28779-fluorine.html
http://www.chemicool.com/elements/fluorine.html
http://www.rsc.org/periodic-table/element/9/fluorine
http://www.chemistryexplained.com/Fe-Ge/Fluorine.html
http://periodic.lanl.gov/9.shtml
http://www.softschools.com/facts/periodic_table/neon_facts/184/
http://chemwiki.ucdavis.edu/Inorganic_Chemistry/Descriptive_Chemistry/Elements_Organized_by_Block/2_p-Block_Elements/Group_18%3A_The_Noble_Gases/Chemistry_of_Neon
http://en.wikipedia.org/wiki/Lithium
http://www.chemguide.co.uk/inorganic/group1/reacth2o.html
http://www.rsc.org/periodic-table/element/3/lithium
http://www.rsc.org/periodic-table/element/4/beryllium
http://www.chemicalelements.com/elements/be.html
http://en.wikipedia.org/wiki/Beryllium
https://www.webelements.com/beryllium/chemistry.html
http://www.britannica.com/EBchecked/topic/62913/beryllium-Be/8230/Compounds
http://education.jlab.org/itselemental/ele004.html
http://www.chemistryexplained.com/elements/A-C/Beryllium.html
http://education.jlab.org/itselemental/ele005.html
http://en.wikipedia.org/wiki/Boron
http://www.chemicool.com/elements/boron.html
http://www.rsc.org/periodic-table/element/5/boron
http://www.livescience.com/28674-boron.html
http://www.encyclopedia.com/topic/boron.aspx
http://www.rsc.org/periodic-table/element/6/crbon
https://www.webelements.com/carbon/
http://education.jlab.org/itselemental/ele006.html
http://en.wikipedia.org/wiki/Carbon
http://www.livescience.com/28698-facts-about-carbon.html
http://www.eoearth.org/view/article/150918/
http://www.lenntech.com/periodic/elements/c.htm
http://periodic.lanl.gov/6.shtml
http://www.encyclopedia.com/topic/carbon.aspx
http://chemwiki.ucdavis.edu/Inorganic_Chemistry/Descriptive_Chemistry/Elements_Organized_by_Block/2_p-Block_Elements/Group_14%3A_The_Carbon_Family/Chemistry_of_Carbon
http://www.britannica.com/science/carbon-chemical-element

http://www.softschools.com/facts/periodic_table/boron_facts/185/
http://www.britannica.com/science/nitrogen
http://www.chemicalelements.com/elements/n.html
http://periodic.lanl.gov/7.shtml
http://www.rsc.org/periodic-table/element/7/nitrogen
https://www.webelements.com/nitrogen/
http://www.lenntech.com/periodic/elements/n.htm
http://education.jlab.org/itselemental/ele007.html
http://en.wikipedia.org/wiki/Nitrogen
http://www.chemistryexplained.com/elements/L-P/Nitrogen.html
https://en.wikipedia.org/wiki/Oxygen
http://www.chemistryexplained.com/elements/L-P/Oxygen.html
http://www.lenntech.com/periodic/elements/o.htm
http://www.webelements.com/oxygen/
http://www.rsc.org/periodic-table/element/8/oxygen
http://periodic.lanl.gov/8.shtml
http://www.britannica.com/science/oxygen
http://periodic.lanl.gov/10.shtml
http://en.wikipedia.org/wiki/Neon
http://www.chemicalelements.com/elements/ne.html
https://www.webelements.com/neon/
http://www.rsc.org/periodic-table/element/10/neon
http://www.chemicool.com/elements/neon.html
http://www.chemistryexplained.com/elements/L-P/Neon.html
http://en.wikipedia.org/wiki/Sodium
http://education.jlab.org/itselemental/ele011.html
http://www.chemicool.com/elements/sodium.html
http://www.livescience.com/28820-sodium.html
http://www.rsc.org/periodic-table/element/11/sodium
http://chemistry.about.com/od/elementfacts/a/10-Sodium-Facts.htm
http://www.webelements.com/sodium/compounds.html
http://chemwiki.ucdavis.edu/Inorganic_Chemistry/Descriptive_Chemistry/s-Block_Elements/Group__1%3A_The_Alkali_Metals/Chemistry_of_Sodium
http://periodic.lanl.gov/11.shtml
http://www.chemistryexplained.com/elements/P-T/Sodium.html
http://www.softschools.com/facts/periodic_table/sodium_facts/187/
http://www.chemistryexplained.com/elements/A-C/Aluminum.html
http://www.livescience.com/28865-aluminum.html
http://www.softschools.com/facts/periodic_table/aluminium_facts/188/
http://periodic.lanl.gov/13.shtml
http://en.wikipedia.org/wiki/Aluminium
http://education.jlab.org/itselemental/ele013.html
http://www.chemicalelements.com/elements/al.html
http://scifun.chem.wisc.edu/chemweek/aluminum/aluminum.html
http://www.rsc.org/periodic-table/element/13/aluminum
http://www.webelements.com/aluminium/chemistry.html
http://en.wikipedia.org/wiki/Phosphorus
http://www.lenntech.com/periodic/elements/p.htm
http://www.rsc.org/periodic-table/element/15/phosphorous
http://www.webelements.com/phosphorus/
http://www.chemicalelements.com/elements/p.html
http://education.jlab.org/itselemental/ele015.html
https://www.organicfacts.net/health-benefits/minerals/health-benefits-of-phosphorus.html
http://www.britannica.com/EBchecked/topic/457568/phosphorus-P
http://environmentalchemistry.com/yogi/periodic/P-compounds.html#Compounds
http://chemwiki.ucdavis.edu/Inorganic_Chemistry/Descriptive_Chemistry/p-Block_Elements/Group_15%3A_The_Nitrogen_Family/Chemistry_of_Phosphorous

http://www.trueorigin.org/atp.php
http://en.wikipedia.org/wiki/Sulfur
http://education.jlab.org/itselemental/ele016.html
http://www.livescience.com/28939-sulfur.html
http://www.rsc.org/periodic-table/element/16/sulfur
http://periodic.lanl.gov/16.shtml
https://www.webelements.com/sulfur/
http://www.chemicalelements.com/elements/s.html
http://www.britannica.com/EBchecked/topic/572661/sulfur-S
http://en.wikipedia.org/wiki/Oxidation_state
http://www.chemicool.com/elements/sulfur.html
http://www.chemistryexplained.com/elements/A-C/Chlorine.html
http://www.chemicool.com/elements/chlorine.html
http://www.webelements.com/chlorine/
http://education.jlab.org/itselemental/ele017.html
http://www.rsc.org/periodic-table/element/17/chlorine
http://www.britannica.com/EBchecked/topic/113561/chlorine-Cl
http://periodic.lanl.gov/17.shtml
http://en.wikipedia.org/wiki/Chlorine
http://environmentalchemistry.com/yogi/periodic/electronegativity.html
http://www.angelo.edu/faculty/kboudrea/periodic/trends_electronegativity.htm
http://en.wikipedia.org/wiki/Argon
http://www.lenntech.com/periodic/elements/ar.htm
https://www.webelements.com/argon/
http://www.livescience.com/29023-argon.html
http://education.jlab.org/itselemental/ele018.html
http://www.rsc.org/periodic-table/element/18/argon
http://www.chemicool.com/elements/argon.html
http://www.chemicalelements.com/elements/ar.html
http://www.britannica.com/EBchecked/topic/33896/argon-Ar
http://www.chemistryexplained.com/A-Ar/Argon.html
http://www.chemistryexplained.com/Pl-Pr/Potassium.html
http://www.britannica.com/EBchecked/topic/472373/potassium-K
http://www.chemicalelements.com/elements/k.html
http://www.chemicool.com/elements/potassium.html
http://www.rsc.org/periodic-table/element/19/potassium
http://education.jlab.org/itselemental/ele019.html
http://www.livescience.com/29025-facts-about-potassium.html
https://www.webelements.com/potassium/
http://en.wikipedia.org/wiki/Potassium
http://en.wikipedia.org/wiki/Calcium
http://www.chemistryexplained.com/elements/A-C/Calcium.html
http://www.britannica.com/EBchecked/topic/88956/calcium-Ca
http://www.chemicalelements.com/elements/ca.html
http://www.chemicool.com/elements/calcium.html
http://www.rsc.org/periodic-table/element/20/calcium
http://education.jlab.org/itselemental/ele020.html
http://www.livescience.com/29070-calcium.html
https://www.webelements.com/calcium/
http://www.lenntech.com/periodic-chart-elements/melting-point.htm
http://www.lenntech.com/periodic-chart-elements/boiling-point.htm
http://en.wikipedia.org/wiki/Scandium
http://www.chemistryexplained.com/Ru-Sp/Scandium.html
http://www.britannica.com/EBchecked/topic/526542/scandium-Sc
http://www.chemicalelements.com/elements/sc.html
http://www.chemicool.com/elements/scandium.html
http://www.rsc.org/periodic-table/element/21/scandium

http://education.jlab.org/itselemental/ele021.html
http://www.livescience.com/29071-scandium.html
https://www.webelements.com/scandium/
http://education.jlab.org/itselemental/ele022.html
https://en.wikipedia.org/wiki/Titanium
http://www.chemicool.com/elements/titanium.html
http://www.chemicalelements.com/elements/ti.html
http://www.rsc.org/periodic-table/element/22/titanium
http://www.chemistryexplained.com/Te-Va/Titanium.html
http://www.livescience.com/29103-titanium.html
https://www.webelements.com/titanium/
http://www.britannica.com/science/titanium
http://www.softschools.com/facts/periodic_table/titanium_facts/193/
http://www.berkeleypoint.com/learning/titanium.html
http://education.jlab.org/itselemental/ele023.html
https://en.wikipedia.org/wiki/Vanadium
http://www.chemicool.com/elements/vanadium.html
http://www.chemicalelements.com/elements/v.html
http://www.rsc.org/periodic-table/element/23/vanadium
http://www.chemistryexplained.com/Te-Va/Vanadium.html
http://www.livescience.com/29155-vanadium.html
https://www.webelements.com/vanadium/
http://www.britannica.com/science/vanadium
http://www.softschools.com/facts/periodic_table/vanadium_facts/200/
http://chemistry.wikia.com/wiki/Vanadium
http://www.globalspec.com/learnmore/materials_chemicals_adhesives/metals_alloys/master_alloys_alloying_additives
http://education.jlab.org/itselemental/ele024.html
https://en.wikipedia.org/wiki/Chromium
http://www.chemicool.com/elements/chromium.html
http://www.chemicalelements.com/elements/cr.html
http://www.rsc.org/periodic-table/element/24/chromium
http://www.chemistryexplained.com/Ce-Co/Chromium.html
http://www.livescience.com/29194-chromium.html
https://www.webelements.com/chromium/
http://www.britannica.com/science/chromium
http://www.softschools.com/facts/periodic_table/chromium_facts/202/
http://chemistry.wikia.com/wiki/Chromium
https://en.wikipedia.org/wiki/Category:Chromium_alloys
http://www.icdacr.com/index.php?option=com_content&view=article&id=91&Itemid=342&lang=en
http://chemwiki.ucdavis.edu/Inorganic_Chemistry/Descriptive_Chemistry/d-Block_Elements/Group_06%3A_Transition_Metals/Chemistry_of_Chromium
http://education.jlab.org/itselemental/ele024.html
https://en.wikipedia.org/wiki/Manganese
http://www.chemicool.com/elements/manganese.html
http://www.chemicalelements.com/elements/mn.html
http://www.rsc.org/periodic-table/element/25/manganese
http://www.chemistryexplained.com/Kr-Ma/Manganese.html
https://www.webelements.com/manganese/
http://www.britannica.com/science/manganese
http://www.softschools.com/facts/periodic_table/manganese_facts/201/
http://chemistry.wikia.com/wiki/Manganese
https://www.americanelements.com/mn.html
http://chemwiki.ucdavis.edu/Inorganic_Chemistry/Descriptive_Chemistry/d-Block_Elements/Group_07%3A_Transition_Metals/Chemistry_of_Manganese
http://education.jlab.org/itselemental/ele025.html
http://education.jlab.org/itselemental/ele026.html

http://www.chemicalelements.com/elements/fe.html
http://www.rsc.org/periodic-table/element/26/iron
http://www.chemicool.com/elements/iron.html
https://en.wikipedia.org/wiki/Iron
http://periodic.lanl.gov/26.shtml
http://www.britannica.com/science/iron-chemical-element
http://www.softschools.com/facts/periodic_table/iron_facts/205/
http://chemistry.wikia.com/wiki/Iron
http://chemwiki.ucdavis.edu/Inorganic_Chemistry/Descriptive_Chemistry/d-Block_Elements/Group_08%3A_Transition_Metals/Chemistry_of_Iron
http://www.webelements.com/iron/compounds.html
https://www.americanelements.com/fe.html
https://www.webelements.com/iron/
https://www.webelements.com/nickel/
http://www.webelements.com/nickel/compounds.html
http://chemwiki.ucdavis.edu/Inorganic_Chemistry/Descriptive_Chemistry/d-Block_Elements/Group_10%3A_Transition_Metals/The_Chemistry_of_Nickel
http://chemistry.wikia.com/wiki/Nickel
http://www.softschools.com/facts/periodic_table/nickel_facts/204/
http://www.britannica.com/science/nickel-chemical-element
http://periodic.lanl.gov/28.shtml
https://en.wikipedia.org/wiki/Nickel
http://www.chemicool.com/elements/nickel.html
http://www.rsc.org/periodic-table/element/28/nickel
http://www.chemicalelements.com/elements/ni.html
http://education.jlab.org/itselemental/ele028.html
http://www.eoearth.org/view/article/154850/
http://www.nickelinstitute.org/NickelUseInSociety/AboutNickel/NickelMetaltheFacts.aspx
http://education.jlab.org/itselemental/ele029.html
http://www.chemicalelements.com/elements/cu.html
https://www.webelements.com/copper/
http://www.webelements.com/copper/compounds.html
http://chemwiki.ucdavis.edu/Inorganic_Chemistry/Descriptive_Chemistry/d-Block_Elements/Group_11%3A_Transition_Metals/The_Chemistry_of_Copper
http://www.softschools.com/facts/periodic_table/copper_facts/208/
http://www.britannica.com/science/copper
http://periodic.lanl.gov/29.shtml
https://en.wikipedia.org/wiki/Copper
http://www.chemicool.com/elements/copper.html
http://www.rsc.org/periodic-table/element/29/copper
http://www.chemistryexplained.com/elements/C-K/Copper.html
https://en.wikipedia.org/wiki/List_of_copper_alloys
https://en.wikipedia.org/wiki/Zinc
http://education.jlab.org/itselemental/ele030.html
http://www.chemicalelements.com/elements/zn.html
https://www.webelements.com/zinc/
http://www.webelements.com/zinc/compounds.html
http://chemwiki.ucdavis.edu/Inorganic_Chemistry/Descriptive_Chemistry/d-Block_Elements/Group_12%3A_Transition_Metals/Chemistry_of_Zinc
http://www.softschools.com/facts/periodic_table/zinc_facts/206/
http://www.britannica.com/science/zinc
http://periodic.lanl.gov/30.shtml
http://www.chemicool.com/elements/zinc.html
http://www.rsc.org/periodic-table/element/30/zinc
http://www.chemistryexplained.com/elements/T-Z/Zinc.html
https://en.wikipedia.org/wiki/Gallium
http://education.jlab.org/itselemental/ele031.html

http://www.chemicalelements.com/elements/ga.html
https://www.webelements.com/gallium/
http://www.webelements.com/gallium/compounds.html
http://chemwiki.ucdavis.edu/Inorganic_Chemistry/Descriptive_Chemistry/p-Block_Elements/Group_13%3A_The_Boron_Family/Gallium
http://www.britannica.com/science/gallium
http://periodic.lanl.gov/31.shtml
http://www.chemicool.com/elements/gallium.html
http://www.rsc.org/periodic-table/element/31/gallium
http://www.chemistryexplained.com/Fe-Ge/Gallium.html
http://www.softschools.com/facts/periodic_table/gallium_facts/207/
http://www.lenntech.com/periodic/elements/ga.htm
http://periodictable.com/Properties/A/ThermalConductivity.v.wt.html
http://www.encyclopedia.com/topic/gallium.aspx
http://www.encyclopedia.com/topic/silicon.aspx
http://www.lenntech.com/periodic/elements/si.htm
https://en.wikipedia.org/wiki/Silicon
http://education.jlab.org/itselemental/ele014.html
http://www.chemicalelements.com/elements/si.html
https://www.webelements.com/silicon/
http://www.webelements.com/silicon/compounds.html
http://chemwiki.ucdavis.edu/Inorganic_Chemistry/Descriptive_Chemistry/Elements_Organized_by_Block/2_p-Block_Elements/Group_14%3A_The_Carbon_Family/Chemistry_of_Silicon
http://www.britannica.com/science/silicon
http://periodic.lanl.gov/14.shtml
http://www.chemicool.com/elements/silicon.html
http://www.rsc.org/periodic-table/element/14/silicon
http://www.chemistryexplained.com/Ru-Sp/Silicon.html
http://www.softschools.com/facts/periodic_table/silicon_facts/189/
http://www.softschools.com/facts/periodic_table/boron_facts/185/
https://www.webelements.com/boron/
http://chemwiki.ucdavis.edu/Inorganic_Chemistry/Descriptive_Chemistry/Elements_Organized_by_Block/2_p-Block_Elements/Group_13%3A_The_Boron_Family/Chemistry_of_Boron
http://periodic.lanl.gov/5.shtml
http://www.foresight.org/Nanomedicine/Ch03_1.html
http://education.jlab.org/qa/mathatom_04.html
http://www.lenntech.com/periodic/elements/cl.htm
https://www.americanelements.com/sc.html
https://www.americanelements.com/ti.html
https://www.americanelements.com/cr.html
https://www.americanelements.com/co.html
https://www.americanelements.com/ni.html
https://www.americanelements.com/cu.html
https://www.americanelements.com/zn.html
https://www.americanelements.com/ga.html
http://www.universetoday.com/46687/black-hole-facts/
http://education.jlab.org/qa/quark_01.html
http://chemistry.about.com/od/elementfacts/a/10-hydrogen-facts.htm
http://www.ducksters.com/science/chemistry/neon.php
http://sciencelearn.org.nz/Contexts/Just-Elemental/Science-Ideas-and-Concepts/How-elements-are-formed
http://scienceline.ucsb.edu/getkey.php?key=24
http://www.livescience.com/22641-cosmic-phosphorus-first-life-astrobiology.html
https://en.wikipedia.org/wiki/Supernova_nucleosynthesis
http://www.scandiummining.com/i/pdf/Scandium-White-PaperEMC-Website-June-2014-.pdf
https://en.wikipedia.org/wiki/Category:Titanium_minerals
https://en.wikipedia.org/wiki/Category:Scandium_minerals

https://en.wikipedia.org/wiki/Category:Calcium_minerals
https://en.wikipedia.org/wiki/Category:Vanadium_minerals
https://en.wikipedia.org/wiki/Category:Potassium_minerals
https://www.esci.umn.edu/courses/1001/minerals/sulfur.shtml
https://www.americanelements.com/vv.html
https://en.wikipedia.org/wiki/Category:Chromium_minerals
https://en.wikipedia.org/wiki/Category:Manganese_minerals
https://en.wikipedia.org/wiki/Category:Iron_minerals
https://en.wikipedia.org/wiki/Category:Cobalt_minerals
https://en.wikipedia.org/wiki/Category:Copper_minerals
https://en.wikipedia.org/wiki/Category:Lithium_minerals
https://en.wikipedia.org/wiki/Category:Beryllium_minerals
https://en.wikipedia.org/wiki/Category:Sodium_minerals
https://en.wikipedia.org/wiki/Category:Magnesium_minerals
https://en.wikipedia.org/wiki/Category:Aluminium_minerals
https://en.wikipedia.org/wiki/Silicate_minerals
https://en.wikipedia.org/wiki/Category:Zinc_minerals
http://webmineral.com/chem/Chem-P.shtml#.VlUwsPnnvIX
http://chemwiki.ucdavis.edu/Inorganic_Chemistry/Descriptive_Chemistry/Elements_Organized_by_Block/2_p-Block_Elements/Group_14%3A_The_Carbon_Family/Chemistry_of_Silicon
https://en.wikipedia.org/wiki/Carbonate_minerals
http://webmineral.com/chemical.shtml#.VltMzvnnvIU
http://webmineral.com/chem/Chem-S.shtml#.VltM4fnnvIU
http://webmineral.com/chem/Chem-N.shtml#.Vlu_-fnnvIU
https://en.wikipedia.org/wiki/Compounds_of_oxygen
http://webmineral.com/chem/Chem-Co.shtml#.Vl0bH_nnvIX
http://www.universetoday.com/101552/earth-is-the-most-exotic-place-in-the-universe/
https://www.astrologers.com/about/history
https://en.wikipedia.org/wiki/Babylonia
https://en.wikipedia.org/wiki/History_of_astrology
https://en.wikipedia.org/wiki/Horoscopic_astrology
http://www.britannica.com/topic/astrology
http://www.nasa.gov/audience/foreducators/5-8/features/F_How_Big_is_Our_Universe.html

Made in United States
Orlando, FL
05 November 2026

90241893R00102